The Excellent Powder
DDT's Political and Scientific History

Donald Roberts & Richard Tren
with Roger Bate & Jennifer Zambone

This edition published by
Dog Ear Publishing
4010 W. 86th Street, Ste H
Indianapolis, IN 46268

www.dogearpublishing.net

dog ear PUBLISHING

ISBN: 978-160844-376-5
This book is printed on acid-free paper.

Printed in the United States of America

“We have discovered many preventives against tropical diseases, and often against the onslaught of insects of all kinds, from lice to mosquitoes and back again. *The excellent DDT powder* which had been fully experimented with and found to yield astonishing results will henceforth be used on a great scale by the British forces in Burma and by the American and Australian forces in the Pacific and India in all theatres.”

Sir Winston Churchill, September 28, 1944 (emphasis added).

Contents

List of Figures

Preface

Modern medicine continues to deliver on its promise of conquest over infectious diseases, yet old infectious diseases continue to rebound and expand in the poorer countries of the world. In our medical arsenal, we have the tools to identify and prevent threats and to treat many health problems we fail to prevent. Possessing these tools, however, will not protect us if we refuse to deploy them, putting ideological and political barriers blindly in the way of the most effective controls. This book discusses one such example of this problem—the environmental crusade that began in the 1960s against the use of DDT and other insecticides in vector control (the preventive measures directed against insects that transmit diseases to humans). Over time, this crusade has shaped the actions of the World Health Organization (WHO) and the agencies and organizations that follow its advice, allowing preventable diseases such as malaria, dengue, and leishmaniasis to infect and kill millions.

DDT is unique in its power to cheaply, effectively, and safely protect poor people in poor countries against diseases. Indeed, because some environmentalists worried about population growth and precisely because of DDT's effectiveness, they became fixated on eliminating it. DDT allowed poor people in developing countries, and the children of poor people, to survive. Leading environmentalists of the time, such as Paul Ehrlich, author of the best-selling book, *The Population Bomb*, railed against growth in human populations and the use of DDT for disease control because they believed that "every life saved this year in a poor country diminishes the quality of life for subsequent generations."[1] Opposition to DDT, as prescribed by Paul Ehrlich, was a means to stop "exported

[1] Hardin, G., "Stalking the wild taboo." http://www.lrainc.com/swtaboo/stalkers/hardin.html (accessed Apr. 11, 2009).

death control."[2] In 1970, *Science* magazine published a paper by self-described ecologist, George Woodwell, whose work we discuss in detail in this book, which offered a solution to his characterizations of environmental pollution: "Fewer people, unpopular but increasing restrictions on technology (making it more and more expensive)."[3] His approach represented a consensus view of the major stakeholders within the environmental movement. European nations and the United States used insecticides to rid themselves of diseases and then pulled up the ladder, denying Africans, Asians, and Latin Americans the benefits of those same insecticides. As a result of these and other environmentalist attacks, DDT was removed from malaria-control programs, costs of malaria control skyrocketed, and the health and welfare of poor people in poor countries plummeted.

Over time, DDT elimination became a global phenomenon. A modern version of this phenomenon and the overwhelming dominance of environmental advocacy over public health was on display at the negotiations of the Stockholm Convention on persistent organic pollutants (of which DDT is one). These negotiations took place over several years ending in 2003. In chapter 7 we present statistics on the composition of those attending the negotiations. The magnitude and inequity of anti-insecticide advocacy compared to representation of the legitimate public health concerns in preserving the use of DDT in developing countries was staggering.

WHO initially supported DDT for malaria control. In 1968, Ehrlich complained, "the World Health Organization . . . refuses to give up DDT for malaria control, claiming that hundreds of millions are doomed without it."[4] History has shown that WHO

---

[2] Ehrlich, P.R., *The Population Bomb* (Cutchogue, NY: Buccaneer Books, 1971), 15.

[3] Woodwell, G.M., "Effects of pollution on the structure and physiology of ecosystems." *Science*, 20 April 1970, 168:429-433

[4] Ehrlich, *The Population Bomb,* 125.

was correct. Yet despite the comprehensive evidence of the efficacy of public-health insecticides, the anti-DDT crusade won. In the face of anti-DDT political pressure, WHO acquiesced and DDT was largely abandoned. Diseases reemerged and millions died.

This book describes some of the ecological and epidemiological principles underlying the public-health use of DDT, a unique chemical with unique actions. No other insecticide matches DDT's broad spectrum of chemical actions or its cost-effectiveness for disease control. The use of DDT in indoor residual spray programs (the spraying of insecticide on the inside walls of houses) produces enormous benefits. When WHO and donors pressured countries to stop their spray programs, enormous increases in disease and loss of life occurred. This book does not simply contrast the public-health benefits of DDT with the rising rates of disease when DDT is not used, however. It also explores how DDT opponents base their opposition on ideology and false claims about DDT's harms to human health and the environment as opposed to truths about the very real, very demonstrable benefits of its use for malaria control.

Another barrier to DDT use was biologist and writer Rachel Carson's crusade against the use of insecticides, which was taken up in the late 1960s and early 1970s by many ecologists and eventually led to the politicized decision by the U.S. Environmental Protection Agency (EPA) to ban DDT. Even today, the ban is propagandized as a major win for the environment and a major loss for the insecticide industry. However, a victory for the environmental movement was not a victory over big insecticide companies. According to Max Sobelman, president of DDT manufacturer Montrose Chemical Corporation, who was intimately involved in defending the use of DDT in the 1970s,

> DDT . . . is an easy mark. It is a $20 million per year business spread over 4 companies—not big business by

> any current yardstick. DDT is not protected by patents. It is being manufactured because there is a demand . . . DDT's biggest boosters and users are the agencies most concerned with human health: The U.S. Public Health Service and the World Health Organization. Between them they buy about 60% of the country's output for world-wide malaria control. Another 20% is for export, and only 20% is used in the U.S.
>
> DDT is fighting for its life today because it always has been a low profit commodity and there is little economic justification for promoting it. So the story of DDT is not widely known, and the opponents of DDT say what they please with little refutation and a press eager to pick up sensational stories from those who seek the spotlight.[5]

The loss of DDT was not a loss to industry, but a loss for public health and for scientific integrity. By the mid-1970s, the annual earnings of just twelve of the major environmental organizations were much larger than the total $20 million of the DDT business (and they are massively larger today).[6]

As the well-known public health scientist and historian Bob Desowitz explains:

---

[5] Sobelman, M. DDT, Selected statements from State of Washington DDT hearings and other related papers (Torrance, CA: DDT producers of the United States, Feb. 1970), 220–21.

[6] A 1977 article on the environmental complex reported that the twelve leading environmental organizations had over $48 million in revenues. Heritage Foundation, "The environmental complex," Nov. 19, 1977, http://www.heritage.org/Research/EnergyandEnvironment/IA4.cfm. In 2002, the NRDC and the EDF (now known as Environmental Defense) each had annual revenues of over $40 million. Arnold, Ron, "The politics of environmental defense," Capital Research Center, Aug. 2004.

> In 1962, there were few defenders of DDT; the agricultural industry actually welcomed the chance to abandon it. It was too cheap, and newer, much more expensive and profitable pesticides were ready for the farmer's market. . . . Giving up DDT made the industry look self-righteous, that they were working in the best interests of humans and nature. There was no dissent, when, in 1972, the newly created Environmental Protection Agency (EPA) banned the use of DDT.[7]

In this book we have described the positions of different interest groups, industries, and individuals. We suppose one could conclude that the message of this book is that environmentalists are generally opposed to DDT and the indoor spraying of insecticides. One might also conclude that environmental toxicologists, environmental epidemiologists, environmental chemists, ecologists, individuals in United Nations (UN) organizations, and members of bilateral and multilateral donor organizations are against spraying, and that only malaria-control specialists favor spraying.

However, even a cursory review of DDT literature will show that there have been environmental scientists, environmental toxicologists, epidemiologists, ecologists, members of the World Health Organization and other UN organizations, and members of bilateral and multilateral organizations who steadfastly supported and defended DDT and other public-health insecticides in disease-control programs. In many cases, their steadfastness was prejudicial to their own careers and put them at risk of being cast in an unfavorable light by reporters. As time has passed, newer generations of defenders have stepped in to carry on the struggle, continuing to argue for the use of DDT and indoor spraying to prevent malaria and other diseases.

---

[7] Desowitz, R.S., *Federal bodysnatchers and the New Guinea Virus* (New York: W.W. Norton, 2002), 64

We have no illusions that this book, or any single book, will turn the tide and wash out the vast propaganda machines' work against public-health insecticides. Decades of well-funded environmental activism have shaped public opinion, and only time and great effort will bring balance and intelligent understanding to this critical matter. Ignorance of the underlying issues breeds the continued successes of those who are willing to sacrifice human health and life for theoretical risks. It is our great hope, however, that this book will help improve understanding and reduce ignorance about DDT and disease-control issues, at least for a few individuals.

Thanks to major new initiatives of DDT advocacy, the U.S. Agency for International Development (USAID) has changed its policies to once again support the use of DDT and the indoor spraying of insecticides in malaria-control programs. These policy changes are promising, but much more needs to be done. The environmental movement continues to work against the use of DDT and other insecticides, and continues to malign those who speak in defense of DDT. On a practical level, the anti-DDT activists may achieve their goal of a DDT-free world simply because there are fewer and fewer producers. The Chinese government, once a major DDT producer, has announced that it will cease production in line with its commitments under the Stockholm Convention. As a result DDT availability is declining and prices are rising, long before any hope of the development of a true alternative and before there has been any real investment in the search for an alternative. Only sustained advocacy will ensure countries' rights to employ the most effective tools available to protect the health and welfare of their neediest citizens.

Increased trade and globalization have brought with them wealth and prosperity, but mosquitoes, parasites, and viruses also travel and find fertile ground in countries that have long forgotten the terrible diseases that once hampered their development and cost millions of lives. Malaria, yellow fever, and dengue once flourished in the United States, and although

America is now wealthier and thus better able to cope with disease epidemics, the threat remains. Without an end to the baseless opposition to insecticide use in public-health programs, the risk of preventable insect-borne diseases in the United States will continue to grow and U.S. citizens may well suffer and possibly die as a result.

The United States should take the lead in reforming public health insecticide policy because it is morally right to prevent countless deaths in poor countries. But if morality is not enough, reforming public-health policy abroad will eventually also save the lives of Americans traveling abroad as tourists, business travelers, or deployed military. It may even save lives at home by reducing the risks of disease introductions in the United States.

The United States makes by far the greatest investment in global public health, funding many disease control and research programs around the world. U.S. taxpayers should insist that their money is spent for maximum effect and that the government should protect this investment by promoting a science-based approach to decisions about public health insecticides. Anything less will undermine this already substantial investment.

We are writing this book for several reasons, principally to set the record straight and to provide the public with another side to the DDT story. This is a story that has been dominated and skewed by erroneous reporting from many in the environmental movement. In our research and writings we have not dodged the thorny and complex issues created by the many claims against DDT. We have compiled extensive observations to enlighten the reader about environmental damage, public health harm, and unique chemical characteristics of DDT. We have also dedicated considerable attention to the national and international politics of DDT. Furthermore we have tackled the uncomfortable but less obvious underlying reasons behind the anti-DDT and, in some cases, the anti-disease control campaigns.

A full and objective understanding of the true DDT story is important. What happened with DDT and the enormous devastation felt by once-effective disease-control programs can happen with other valuable chemicals and technologies that drive human progress. At the end of this book we make specific recommendations for changes in the regulatory control of public health insecticides, changes that might in the long run bring balance and equity to a whole span of public health endeavors.

Acknowledgments

Today, there are many knowledgeable people who understand that there are great disparities between the claims against insecticides and the reality of lost health and life as a result of not using them. These individuals include lawyers, medical doctors, professional writers, philosophers, politicians, scientists (including several Nobel laureates), and the lay public. Their signing of the petition against the DDT ban as first proposed at the Stockholm Convention on Persistent Organic Pollutants demonstrates their understanding of the issue. To name only a few of these individuals would be a disservice to the many who have refused to be misinformed and who have remained steadfast in defense of those who have been subjected to the ever-increasing risks of disease and death and of the millions who have unnecessarily sickened or succumbed to insect-borne diseases.

Of those who have contributed to the preparation of this book, we would like to specifically acknowledge the hard work of Ben Schwab, Lorraine Mooney, Nick Schulz, Kim Hess, Amy Fontinelle, Joanna Andreasson, Cheryl Jackson, Raleen Bagg and Jasson Urbach. We also thank Professor Mark Roseman for his contributions to an early draft of text on organochlorines and lipid physiology. Those who have, in one way or another, assisted in preparation of text are not responsible for any errors or mistakes that we might have made in this book. Lastly we would like to acknowledge Johnette Roberts for her patience and understanding over the years of research and iterations of this book.

Chapter 1

Chemicals and Life

For such a controversial chemical, DDT had an unimpressive beginning. It was created and then forgotten for sixty-five years. In 1874, a German graduate student, Othmar Zeidler, was attempting to generate new substances as part of his thesis. When he reacted chloral hydrate with chlorobenzene in the presence of sulfuric acid, he was simply happy to discover that dichloro-diphenyl-trichloroethane was the result. He wasn't interested in any of its properties, merely its existence. He wrote up the result of his experiment and moved on.[8] DDT had to wait until the 1930s before Dr. Paul Mueller discovered its life-saving properties.

Mueller worked for the Swiss chemical company J. R. Geigy as a research chemist. In the mid-1930s, the company was interested in finding a chemical that would control clothes moths. Mueller was assigned to the clothes moth project and tested a variety of substances, eventually turning his attention to chlorinated organic compounds.

Mueller synthesized DDT and found that when he applied it to the test flies that he kept in glass boxes, they died. Mueller noted the success, cleaned the box thoroughly, added new flies, and then observed something quite peculiar. Although he had not treated the box with any new potential pesticide, the new flies also died. Intrigued and excited by this result, Mueller tested DDT further and found that it also killed the Colorado potato beetles that were becoming a scourge on Switzerland's crops. In 1939, Mueller applied to the Swiss Patent Office for a

[8] Zimmerman, O.T. and Lavine, I., *DDT, killer of killers* (Dover, NH: Industrial Research Service, 1946), 30–31. Fans of the gritty gumshoe genre might better recognize choral hydrate as a "Mickey Finn."

patent, and the Swiss government applied the pesticide to its potato crops and had the satisfaction of seeing the potato beetles quickly exterminated.[9]

At the time the Swiss were killing potato beetles, the rest of Europe was killing each other in World War II. Combat is not the sole killer during war. Diseases, including insect-borne ones, flourish in war, decimating populations. One such disease is typhus. Spread by body lice that thrive in the unsanitary, crowded conditions of war, typhus has been a fixture of war for centuries. It killed thousands of soldiers, civilians, and prisoners during World War II, including diarist Anne Frank.

After the Swiss saw how thoroughly DDT squashed the potato beetle infestation, they began testing its effects on other insects. By 1942, the Swiss had determined that DDT killed flea beetles, apple-blossom weevils, apple sawflies, raspberry beetles, cabbage moths, and grain weevils. Extensive tests with DDT in schools and refugee camps found that DDT killed fleas, mites, and lice. Testing in barns and barracks showed that flies also succumbed to DDT. Realizing the power of this new

---

[9] Ibid., 31–32; In a reflection of the life-saving properties of DDT as an insecticide, Mueller won the Nobel Prize for Physiology or Medicine for his discovery in 1948. In his speech, Mueller listed the criteria he sought in the "ideal insecticide": (1) great insect toxicity, (2) rapid onset of toxic action, (3) little or no mammalian or plant toxicity, (4) no irritant effect and no or only a faint odour (in any case not an unpleasant one), (5) the range of action should be as wide as possible, and cover as many Arthropoda as possible, (6) long, persistent action (i.e., good chemical stability), (7) low price. DDT met all of these criteria but the second. DDT is not a knock-down poison like pyrethrum, but rather possesses a slow-acting toxicity, killing in a matter of hours or days instead of instantaneously. This is, said Mueller, "the reason why initially the biologists were not very interested in the substance; pyrethrum and rotenone had accustomed them to expect a rapid knockdown and they did not understand that long residual activity far outweighed the slow toxic process." Mueller, P.H., "Dichloro-diphenyl-trichloroethane and newer insecticides," Nobel lecture, Dec. 11, 1948.

insecticide, and in keeping with their neutrality, the Swiss released the formula, samples, and data to both the Axis and the Allies in 1942.[10]

The Germans ignored the information, but the data greatly impressed the Allies, particularly those relating to DDT's toxicity to body lice. However, toxicity was not DDT's most important characteristic. Other chemicals were more toxic—but they were also toxic to humans. Nicotine for example is a natural product that has been used as an agricultural insecticide. It is more toxic to insects than DDT. Unfortunately, it is also violently toxic to humans and would never do for a use that involved human exposure. In contrast, DDT was toxic to insects at doses that were safe for humans. That it was active for long periods after application was a bonus.

Even though the Swiss had already conducted extensive toxicity tests, both the Americans and the British decided to conduct their own tests on the pesticide.[11] When their tests confirmed the Swiss results, both the Americans and the British realized that DDT could quell the deadly outbreaks of typhus that ripped through war- and famine-weakened populations and killed a quarter of the people infected.[12] Their chemical industries swung into action. By late 1943, J. R. Geigy's subsidiary in the United States, Cincinnati Chemical Works, was producing massive quantities of DDT. By early 1944, fourteen other American chemical companies were also producing DDT, as were British corporations.[13]

---

[10] Zimmerman and Lavine, *DDT*, 32–33.

[11] West, T.F. and Campbell, G.A., *DDT and newer persistent insecticides* (New York: Chemical Publishing Co., 1952), 2.

[12] Spielman, A. and D'Antonio, M., *Mosquito: the story of man's deadliest foe* (New York: Hyperion, 2001), 116–26.

[13] A fascinating and instructive detail of this production was Geigy's handling of DDT patents. Geigy, of course, had patented DDT in several countries including Switzerland, the United States, Canada, and Great Britain, but recognizing DDT's value to the world, Geigy

Production couldn't have happened at a better time. Beginning in October 1943, typhus broke out in Naples, Italy.[14] It soon reached epidemic proportions. The U.S. Army took steps to control the outbreak using what pesticides it had, but the measures weren't effective until the Army started using a pesticide that it had initially tested on prisoners of war in North Africa, the DDT-containing Neocide. The military mixed Neocide with an inert powder and dusted it on troops and refugees with great effect. In January 1944, two delousing stations dusted 1.3 million civilians. (One station handled 72,000 people in one day.) Within three weeks of the dusting, the

exercised a very liberal licensing policy. In an act of generosity, they asked manufacturers to pay only a nominal fee to produce insecticides that contained DDT. The fee was "remarkably low when one considers the tremendous cost of the many years of development work by this company." Zimmerman and Lavine, *DDT*, 39.

[14] The first cases of typhus in Naples were recorded in March 1943 among Italian solders returning from the Russian front. These patients were admitted and treated in the military hospital; however, it wasn't until the German soldiers departed in September 1943 that the incidence of the disease could escalate through the city. The hygiene conditions at Poggioreale prison were practically non-existent after the bathing facilities were demolished as a result of Allied air raids. Inmates were almost all louse-infested and were crowded together, especially during air raids. On September 23, 1943, the Germans ordered the evacuation of some 300,000 residing in the zone around the port and coastal area, forcing them to live in crowded areas elsewhere in the city. On September 30, 1943, shortly before the Germans departed, they opened the doors of Poggioreale prison, releasing louse-infested prisoners into the already crowded and unhygienic areas of Naples. As a result, typhus spread rapidly, with a total of approximately 2,020 cases and 429 deaths from July 1, 1943 to June 1, 1944. This should be seen as a conservative estimate as many cases were not reported. Soper, F.L., Davis, W.A., Markham, F.S., and Riehl, L.A., "Typhus fever in Italy, 1943–1945, and its control with louse powder," *Am Journ Hyg*, vol. 45, no. 3 (May 1947), 305–34.

epidemic was under control, with the number of civilian cases falling from 305 in the peak week to 155 in the next.[15]

The delousing of concentration camp prisoners was just as impressive. One survivor of Belsen concentration camp recalls:

> After two-three days at the hospital, we have our first encounter with the pesticide DDT. When the English soldiers enter the hospital room with sprayers filled with this product, we all look at them with contemptuous superiority. They're planning on using this puny white powder to destroy all these millions of lice! But hundreds of the most radically conducted Entlausungs [delousing chambers that ineffectively used heat to kill body lice in clothing (Bau, 2008)] hadn't helped. Yet, right in front of our eyes, something close to a miracle starts to happen! Slowly, the incessant itching, so painful on our puss infected, ulcerated skin, starts to vanish, and this great relief finally convinces us that we have really been liberated. O Great, Powerful Benefactor, Inventor of the White Powder![16]

It was an astonishing achievement. DDT had defeated typhus, which had been a scourge of war-torn areas for millennia. So impressive was the achievement that in 1948, Mueller received the Nobel Prize for Physiology or Medicine. In his introductory speech for the prize, Professor G. Fischer stressed what an extraordinary accomplishment it was for Mueller to have discovered a pesticide that could stop the spread of such a deadly disease.

> Towards the end of the Second World War, typhus suddenly appeared anew. All over the world research

[15] West and Campbell, *DDT*, 8.

[16] Baumslag, N. 2005. *Murderous medicine: Nazi doctors, human experimentation, and typhus.* (Praeger Publishers, Westport, CT), 27.

> workers applied their energies to trying to discover an effective delousing method. Results, however, were not very encouraging. In this situation, so critical for all of us, deliverance came. Unexpectedly, dramatically presented out of the blue, DDT appeared as a deus ex machina.
>
> In October of 1943 a heavy outbreak of typhus occurred in Naples and the customary relief measures proved totally inadequate . . . in a period of three weeks the typhus epidemic was completely mastered. Thus, for the first time in history a typhus outbreak was brought under control in winter. DDT had passed its ordeal by fire with flying colors.
>
> Since those days DDT has been used in large quantities in the evacuation of concentration camps, of prisoners and deportees. Without any doubt, the material has already preserved the life and health of hundreds of thousands.[17]

DDT's success in controlling typhus was based in part on its persistence. DDT took a long time to kill the flies that had died in Mueller's lab in the cleaned glass box, but then it continued to kill long after its application.[18] This slow killing action and

---

[17] Fischer, G., Presentation speech for the Nobel Prize in physiology or medicine, 1948, http://www.nobel.se/medicine/laureates/1948/press.html (accessed Apr. 10, 2009).

[18] "In 1945, when DDT was still relatively untried, I met an English Major in Germany who told me he had treated the windowpane of his room with DDT since he was plagued by masses of flies. After the DDT solution had been sprayed on, the flies died and lay in heaps on the window ledge. The following morning a German soldier entered and thoroughly cleaned the window. When the Major noticed this he couldn't help crying, 'Goodbye my DDT!' But this farewell was uncalled-for. In spite of the thorough cleaning, the windowpane retained its deadly action on the flies. This little story amply illustrates how persistent DDT is and how small the dosage required." Fischer, G., Presentation Speech for the Nobel Prize.

persistence were unusual and made DDT unlike other chemicals used in pest control. Mueller initially considered DDT's persistence a flaw, but in fact, it turned out to be DDT's great virtue. One application of the pesticide killed the lice not for a week, as previous pesticides had, but for months, an inestimable benefit as it prevented typhus epidemics from starting, much less spreading through the population or being carried to other areas. The military was so impressed by the effectiveness of the pesticide in killing body lice that it started using it for the control of another insect-borne disease: malaria.[19] During World War II, malaria plagued troops in Europe, the Pacific, Burma, and Malaysia. While drugs were available to treat the infection, malaria was still a debilitating disease.[20] The authorities suspected that the characteristics that made DDT attractive in controlling typhus would also make it effective in controlling malaria. Even Winston Churchill enthused in a radio broadcast on September 28, 1944,

> We have discovered many preventives against tropical diseases, and often against the onslaught of insects of all kinds, from lice to mosquitoes and back again. The excellent DDT powder which had been fully experimented with and found to yield astonishing results will henceforth be used on a great scale by the British

[19]DDT was especially welcome for the control of malaria and typhus because two of the chemicals traditionally used, pyrethrum and derris, were in short supply. The Japanese apparently had extensive control of the pyrethrum flowers, and when the Axis captured British and Netherlands East Indies, it cut off the main supply of derris. West and Campbell, *DDT*, 9.

[20] "At any given time, one-third of his [Gen. Douglas MacArthur's] fighting men were in the throes of malaria, one-third were recovering and only the final third were truly fit for combat. If the problem [malaria] were to persist, said the general, the war would be a very long conflict." Spielman and D'Antonio, *Mosquito*, 142.

> forces in Burma and by the American and Australian forces in the Pacific and India in all theatres.[21]

DDT would soon surpass all expectations in controlling malaria around the world, and the "excellent" powder would save millions of lives.

DDT's remarkable success in controlling insects that spread disease did not go unnoticed to farmers, who battle a wide range of insects and other pests as they struggle to produce food. DDT was soon used in agriculture and in gardens. It was sprayed aerially by planes, on ground from vehicles, and out of small domestic spray cans. DDT use in agriculture successfully controlled a wide range of agricultural pests and allowed for the increased production of food.

DDT Chemistry

Following the discovery of DDT's insecticidal properties, it became both heavily used and intensively studied. Through many decades, fearful claims about its possible adverse effects on human health and on wildlife have dominated DDT coverage in the popular press. The fearmongers' concerns did not prove true, though. A second look at DDT chemistry and the chemistry of natural chemicals and processes helps explain why fearful projections of DDT harm were false.

DDT is relatively easy to produce, its patent was never enforced, and it is cheap compared to modern insecticides. DDT is a few times more toxic than table salt,[22] but far less toxic than

[21] Quoted in West and Campbell, *DDT*, 11.

[22] Bennett, S.M., "Insecticides: DDT," http://www.the-piedpiper.co.uk/th13(l).htm (accessed Apr. 10, 2009); The Salt Institute, "Salt FAQs," http://www.saltinstitute.org/About-salt/Salt-FAQs (accessed Apr. 10, 2009).

nicotine.[23] There has been no known human death or illness from an environmental exposure to it.

DDT is not soluble in water, so it cannot be sprayed as a water solution. When DDT was first used for malaria control in the mid-1940s, it was sprayed on walls as a kerosene solution. Those early uses of DDT were expensive and the smell of kerosene was objectionable. Later, an organic solvent was used to create a DDT solution at the factory. The DDT solution was used to coat particles of talc because talc is inert, cheap, and easy to use. The talc particles are also treated with a dispersing agent. Treated in this way, the particles will readily disperse in water, resulting in a uniform suspension of DDT-laden talc in water, which can then be sprayed on walls. A fine, whitish-gray film remains once the walls dry.

The change from a kerosene formulation to a water-dispersible powder formulation was a giant step forward. Spray teams no longer needed to transport large quantities of kerosene. They carried small sachets of dry powder, pre-weighed for a suspension in water to fill a five gallon spray tank. Teams could use water from local streams, prepare the suspension, and spray.

DDT has low vapor pressure, so people can work with the chemical and not acquire toxic quantities from the air. Depending on how it is formulated, DDT is often not readily absorbed through skin, so people can get DDT on them and not be harmed. Research has shown that even when inhaled on dust particles, the natural defenses of lung tissues move the particles back out of the lungs, where the particles are then swallowed. Ingestion is the primary way humans acquire DDT, but the quantities obtained from inhalation are too small to be harmful. Human toxicity of DDT will be discussed later.

---

23 International Program on Chemical Safety, "Nicotine," http://www.inchem.org/documents/pims/chemical/nicotine.htm#SubSectionTitle:7.2.2%20Relevant%20animal%20data (accessed Apr. 10, 2009).

As stated on a popular website, in reference to an individual human, "you are chemistry—all life develops from and consists of (bio)chemical processes."[24] At least thirty thousand chemicals are known to be present in humans. In the big scheme of natural and man-made chemicals, DDT is a very simple and easily produced compound. Once DDT is actually used in human activities, such as malaria control, it is managed by the natural world through mechanisms of degradation, transformation, or storage that are employed within the environment and by living organisms to manage a huge variety of natural chemicals.

DDT is one of many chemicals that will not dissolve in water, but will dissolve in fat and other carbon-based solvents, like oil and kerosene. DDT is an organic molecule, meaning that it contains carbon atoms. Salt and water are inorganic molecules—they do not contain carbon. Within the lipid (fatty) fraction of human blood are a great variety of fat-soluble molecules, and within the water fraction are a great variety of water-soluble molecules. Many of these fat- and water-soluble molecules are essential for life. Whether one of these molecules is harmful, neutral, or helpful to a living organism is often just a factor of concentration. Even when a harmful concentration occurs, a living organism can often act upon the molecule by breaking it down, transforming it into something less harmful, or temporarily storing the chemical in a form that is not biologically active. The metabolism of a living organism includes a great diversity of pathways for creating and destroying chemicals.

Great harm can result from sudden exposure to overwhelming concentrations of even relatively innocuous molecules. Man's recreational use of alcohol illustrates the issues of dose, time, and toxicity. Ethyl alcohol, or drinking alcohol, is a product of natural fermentation. It is not very toxic when taken in moderation, in large part because it is quickly metabolized to carbon dioxide and water, both of which are readily eliminated

[24] Chemicals for Life, www.chemforlife.org, (accessed Sept, 19, 2009).

by the lungs and kidneys. One must consume considerable amounts to get the toxic effects of drunkenness and a hangover. Yet, death can occur if a large amount of alcohol is ingested quickly. Humans consume a great amount of alcohol, but death is rare because alcohol is not very toxic.

In contrast, there are other molecules that are highly toxic and very dangerous. Biological toxins that are synthesized by living organisms, such as botulinum toxin, can be immensely dangerous. Many of these natural toxins are hundreds of millions of times more toxic than alcohol or DDT. The body's violent response to small doses of botulinum toxin and other natural toxins reflects their extreme biological activity and toxicity. Nevertheless, even these highly toxic chemicals can be used in low concentrations for constructive purposes. BOTOX® is one example.

Living organisms have processes and requirements for acquiring and metabolizing a wide range of chemicals. Consider the chemical cholecalciferol. Like DDT, it is a non-polar organic compound, but it is somewhat more toxic.[25] Cholecalciferol is formulated as a rodenticide and used to kill mice and rats, yet most people know this chemical as vitamin D. Vitamin D is an essential, fat-soluble vitamin. It can be stored in body fat, a process sometimes referred to as compartmentalization, in very large amounts, but beyond a certain threshold level, vitamin D will be metabolized and eliminated from the body. In a similar fashion, the fat-soluble molecule DDT will not accumulate, that is to say, be compartmentalized in fat cells, in the human body indefinitely. Beyond a certain threshold level, it too is metabolized and eliminated from the body.[26]

---

[25] Science Lab.com, "Material safety data sheet, cholecalciferol," http://www.sciencelab.com/xMSDS-Cholecalciferol-9923455 (accessed Apr. 11, 2009).

[26] In long-term dosing studies with human volunteers, only limited quantities of DDT are stored in fat and the quantities will not go above a certain level even with continued daily dosing. This level will vary

Environmentalists and DDT

Environmentalists have had great success with campaigns to scare people because DDT can accumulate in body fat. They claim it will eventually cause grievous harm, and they aggressively project false claims that natural chemicals do not have this characteristic. Common themes in environmentalist opposition to DDT are claims that it cannot be degraded, that it accumulates up the food chain, and that it is dangerously toxic. Environmentalists have successfully promoted their superstitious beliefs that we must act against the production and use of the chemicals they oppose in order to ensure our very survival. But there is no scientific merit to their claims. Environmentalists use their claims to demonize DDT—to portray it as profoundly alien to the natural world. The truth is that many natural chemicals exhibit the same characteristics as DDT, and some are so toxic as to make DDT seem as harmless as the talc it binds to.[27] Ciguatoxin, for example, is a fat-soluble organic compound found in marine ecosystems.[28] It is produced by dinoflagellates, which are single-cell plants that are smaller than a needle's point.

---

with the dose given, but a plateau will be reached beyond which there is no additional DDT storage. Additionally, once dosing ends, the quantity of DDT in fat begins to decline, and the rate of decline is greater with greater quantities in storage. For highest storage, only 32–35 percent remained two years later. None of the volunteers showed an illness attributable to DDT, not even those who received 35mg per day. See Hayes, W.J., Jr., "The degree of hazard to man of DDT as used against malaria" (Geneva: World Health Organization [WHO], 1971), WHO/VBC/71.251.

[27] Vetter, W., "Marine halogenated natural products of environmental relevance." *Rev Environ Contam Tox*, vol. 188 (2006): 1–57.

[28] Hung, Y.M., et. al., "Persistent bradycardia caused by ciguatoxin poisoning after barracuda fish eggs ingestion in Southern Taiwan," *Am J Trop Med Hyg*, vol. 73, no. 6 (2005):1026–27.

See also North West Fisheries Science Center, "Harmful Algal Blooms," http://www.nwfsc.noaa.gov/hab/habs_toxins/marine_biotoxins/index.html (accessed Apr. 11, 2009).

Ciguatoxin accumulates in the food chain and large, predacious fish accumulate it in dangerous quantities. The toxin does not harm the fish, but as little as 0.1 microgram of ciguatoxin per kilogram of body weight is enough to produce illness in humans. This molecule is hundreds of thousand of times more toxic than DDT, and an estimated 50,000 cases of poisoning occur each year, with many cases occurring in the United States. Vitamin A is also present in very high concentrations in some animal species at the top of the food chain, and people have suffered vitamin A toxicity by eating the liver tissue of animals such as polar bears, sharks, and groupers.[29]

The beginning years of DDT use, in the 1940s, were exciting times of exploration and learning. During those years, DDT was used on all manner of food, including in fruit orchards, on vegetable farms, in cattle pastures, in chicken feed, and in kitchens for fly control. According to the U.S. Environmental Protection Agency (EPA), around 675,000 tons of DDT were used in the United States alone between the 1940s and early 1970s. In the peak year of DDT use, 1959, almost 80 million pounds of DDT were applied.[30] With extensive and intensive DDT use in agriculture, DDT appeared in animal fat. Lard, as a product of animal fat, would have further amplified human exposures to DDT. Yet, those years of higher DDT exposures were marked with continual improvements in human health, wealth, and longevity.[31] Through it all, not one single human death or human illness was shown to be a direct result of environmental exposures to DDT or to DDT accumulations in the food chain.

---

[29] Physical and Chemical Theoretical Laboratory, Oxford University, "Safety (MSDS) data for Vitamin A," http://ptcl.chem.ox.ac.uk /MSDS/VI/vitamin_A.html (accessed Apr. 11, 2009).

[30] U.S. Environmental Protection Agency, "DDT ban takes effect," Dec. 31, 1972, http://www.epa.gov/history/topics/ddt/01.htm (accessed Apr. 15, 2009).

[31] Goklany, I., *The improving state of the world.* (Washington, D.C.: Cato Institute, 2007), 19-55

DDT can be metabolized and excreted. So, while humans compartmentalize it in body fat, there are also well-established mechanisms for DDT removal and degradation. Toxicologists Donald Morgan and Clifford Roan precisely and quantitatively defined the pathways and processes for DDT storage and degradation in 1972.[32] They demonstrated human capacity to excrete ingested DDT at high rates. In other words, they showed that humans are well equipped to rapidly eliminate DDT. Their work was preceded by the studies of Wayland Hayes, a toxicologist with the U.S. Department of Health Education and Welfare, in 1958,[33] 1965,[34] and 1971.[35] The primary degradation and storage product of DDT in humans is a molecule known as DDE, and the primary excretory product from DDT degradation is DDA. DDT is only a few times more toxic than table salt, and DDE is many times less toxic than DDT. DDA is only slightly more toxic than DDE.

Double Standards of Environmental Activism Against DDT

DDT was an early target of the environmental movement. Over the past forty-five years, environmentalists have maintained a broad campaign against man-made chlorinated hydrocarbons, especially DDT. They claim that their advocacy

[32] Morgan, D.P. and Roan, C.C., "Loss of DDT from storage in human body fat." *Nature*, vol. 238 (July 28, 1972): 221–23.

[33] Hayes, W.J., Jr., et al., "Storage of DDT and DDE in people with different degrees of exposure to DDT," *AMA Arch Ind Health*, vol. 18, no. 5 (1958): 398–406.

[34] Hayes, W.J., Jr., "Review of the metabolism of chlorinated hydrocarbon insecticides especially in mammals," *Ann Rev Pharmacol*, vol. 5 (1965): 27–52.

[35] Hayes, W.J., Jr., et al., "Evidence of safety of long-term high oral doses of DDT for man," *Arch Environ Health*, vol. 22, no. 1 (1971): 119–36.

against chlorinated hydrocarbons is to protect human health. However, they target their campaign against only chlorinated hydrocarbons applied in the environment.

For decades, chloroquine has been the primary drug for treating and preventing malaria. Chloroquine is consumed, whereas DDT is not consumed, only sprayed on house walls. Chloroquine is man-made (that is, synthetic), and less acutely (immediately) toxic than DDT, but its oral toxicity is broadly similar to DDT. It seems inconsistent that researchers do not preoccupy themselves with the potential for human harm caused by chloroquine as they do with DDT since chloroquine is also a chlorinated hydrocarbon. In animal studies, chloroquine, unlike DDT, shows actions that interfere with normal embryonic development and causes mutations, and over the decades, vast quantities of chloroquine have been consumed.[36] In 2003 alone, Mexico distributed seven million chloroquine tablets, and Nicaragua dispensed a similar number. Chloroquine has proven to be relatively safe for humans, but not entirely.[37] On the other hand, based on historical data and on how DDT is used in malaria control, DDT should be classified as entirely safe for humans.

Because of the similarities between the two, it seems anti-DDT activists would be as concerned about chloroquine as they are about DDT if they were truly concerned about the impact of man-made organochlorines on human health. History has proven that chloroquine is a wonderful drug for treating malaria cases. History has also shown that DDT is a wonderful chemical for preventing malaria cases.

---

[36] Physical and Chemical Theoretical Laboratory, Oxford University, "Safety (MSDS) data for Vitamin A," http://ptcl.chem.ox.ac.uk/MSDS/VI/vitamin_A.html (accessed Apr. 11, 2009).

[37] Crouzette, J., et al., "Experimental assessment of the protective activity of diazepam on the acute toxicity of chloroquine," *J Tox Clin Tox*, vol. 20, no. 3 (May 1983): 271–79.

Summary

In this chapter, we have described some of the basic properties of DDT. DDT is not soluble in water, but will dissolve in fats and will be stored in body fat. Once in the human body, above a certain level, DDT will be metabolized and excreted.

Many people fear DDT and other man-made chemicals simply because they have been created by man. But there is nothing inherently dangerous or unnatural about chemical compounds such as DDT. Furthermore, DDT is generally less dangerous to humans than any number of natural chemicals, including some vitamins and medicines that we consume without a second thought. In malaria control, DDT is not even consumed, but is sprayed in a water-based suspension on the inside walls of houses. Despite a lack of evidence, environmentalists fear that DDT's persistence, one of its great assets in malaria control, will cause harm to human health and the environment. As we describe in subsequent chapters, these fears are misplaced and the campaigns against DDT have done great harm to human health.

Chapter 2

Malaria—Destroyer of Economies, Families, Health, and Lives

President George W. Bush mentioned malaria twice in his State of the Union speech on January 31, 2006. This was likely a surprise for most viewers in the United States, where malaria was defeated as an endemic disease more than fifty years ago. Those of us who live in advanced, wealthy countries might want to ignore malaria and other such diseases. And why shouldn't we? We do not suffer from them, and our governments provide considerable financial support to UN organizations like the World Health Organization (WHO), the Children's Fund (UNICEF), and others. Our financial resources and the organizations our public funds support help combat malaria and other diseases. So our duty is done—we should feel no further obligation to poor people in poor countries. Or is there more to this story than we think?

In truth, while we have good intentions and some of our resources are even well targeted, our aid comes with baggage. Our environmental concerns and policies inflict indirect and occasionally direct harm on hundreds of millions of poor people in poor countries. The failure of the organizations we support and the historic failure of our foreign assistance are breathtaking in scope.

WHO is the United Nations' specialized agency for health. It is governed by the 192 member states that make up WHO, and every year those states meet in order to approve WHO's program and budget at the World Health Assembly (WHA).

WHO is the foremost UN organization responsible for malaria control. The World Health Assembly is the ultimate decision-making body for WHO. The WHA adopted a resolution in 1997 calling on all countries to reduce the use of insecticides for disease control.[38] DDT was carefully identified in the resolution as one that should be phased out of use. Not mentioned, even as a footnote, was the fact that there were no broadly applicable and effective means of controlling malaria without the use of insecticides, and DDT in particular (a situation that remains to this day). So how can we reconcile the 1997 resolution with a public-health need at a time when developing countries were suffering from a global pandemic of dengue fever and malaria was reemerging? The resolution was an unthinking (but by some, a flagrant) endorsement of environmental extremism over public health. Worst of all, this anti-public-health resolution was endorsed by the world's premier health policy-making body.

Mosquito Biology

In order to understand how politics has contributed to the resurgence and reemergence of malaria, it is important to understand the various methods of malaria control and the politics that govern that control. For that, a grasp of mosquito biology and the basic relationship between man, mosquito, and malaria is required.

Mosquitoes are two-winged flies of which there are over 2,500 different species. Almost every terrestrial location will have many different species of resident mosquitoes. Most men never tire of hearing that the females are more dangerous than the males of the species. It's true: only female mosquitoes feed on blood, and thus only female mosquitoes transmit disease. Females need the protein in blood to produce eggs. Male mosquitoes, free of parental responsibility, feed mainly on nectar.

[38] WHO, 50th World Health Assembly, Resolution WHA 50.13, (Geneva, May 1997).

How, when, and where the female lays her eggs and gets her meals of blood are genetically programmed according to her species. So highly specific is this programming to each species that it affects methods of mosquito control. One commonality between species, however, is water. Mosquitoes must undergo their larval and pupal stages in water, so female mosquitoes lay their eggs on, in, or near water. However, the water that the female mosquito chooses for laying her eggs varies vastly from species to species. Some, such as *Aedes aegypti*, the primary vector (the word vector is used to label an insect that bites and passes disease agents to humans and other animals) of the dengue and yellow fever viruses, can lay their eggs in small containers, such as cans and old tires. On the other hand, South America's major malaria vector, *Anopheles darlingi*, chooses large water bodies, such as rivers and streams, in which to lay her eggs.[39] Some species, such as *Culex pipiens*, which transmits St. Louis encephalitis and West Nile Virus, do not seem to mind if the water body is highly polluted and associated with human activity.

While each mosquito species is tied to specific types of water, some species are less restricted to specific habitats than others. *An. albimanus* is a ubiquitous species that will lay its eggs in a wide variety of waters.[40] In northern Belize, the larvae of this malaria vector commonly inhabit reed marshes that have floating mats of cyanobacteria, commonly known as blue-green algae.[41] In regions without reed marshes, however, this mosquito lays eggs in a variety of grassy, sun-exposed pools of rainwater and will even use rice fields as a breeding ground. In contrast,

---

[39] Manguin, S. et. al., "Characterization of *Anopheles darlingi* (Diptera: Culicidae) larval habitats in Belize, Central America," *J Med Entomol*, vol. 33, no. 2 (1995): 205–11.

[40] Rejmankova, E. et. al., "Environmental and regional determinants of *Anopheles* (Diptera: Culicidae) larval distribution in Belize, Central America." *Environ Entomol*, vol. 22, no. 5 (1993): 978–92.

[41] Rejmankova, E. et.al., "Predictions of adult *Anopheles albimanus* densities in villages based on distances to remotely sensed larval habitats," *Am J Trop Med Hyg*, vol. 53, no. 5 (1995): 482–88.

another mosquito species, *An. vestitipennis*, found in the same regions in Belize, is much pickier. It lays its eggs almost exclusively in dense, tall macrophyte (aquatic plants visible to the naked eye) marshes, such as cattail marshes.[42] The more strongly a mosquito species links to a specific type of habitat, the more likely it is that the absence of that habitat will indicate the absence of that mosquito.

In the water, immature mosquitoes, or larvae, emerge from eggs and undergo rapid development. Under optimal warm conditions and adequate nutrition, the larvae can complete the four stages of larval development within a week, after which they transform into pupae. During the pupal stage, which can last as little as two days, the mosquitoes do not feed, but transform themselves from larvae into winged adults. Once the fully formed adult is under the pupal skin, the pupa floats up to the water surface. The skin splits in the middle and the winged adult pushes out and rests on the surface. The adult mosquito then uses its muscles and the (hydrostatic) pressure of its internal fluids to unfurl its wings and other appendages, and waits for its exoskeleton to harden. The mosquito then flies to a resting place, usually in vegetation, and after as little as twenty-four hours, mating can begin. Within two days of emerging from water, females may seek their first blood meal.

Adult female mosquitoes have a specific mission of laying fertile eggs. To fulfill this mission, the female must mate and feed. The sexes come together briefly for mating. Either the males and the females swarm around each other and mate, or the males seek out the preferred host and mate with a female just before she takes a blood meal from the host.

These behaviors do not mean that mosquitoes think. Mosquitoes do not process information and then choose between

---

[42] Rejmankova, E. et.al., " Characterization and detection of *Anopheles vestitipennis* and *Anopheles punctimacula* (Diptera: Culicidae) larval habitats in Belize with field survey and SPOT satellite imagery," *J Vect Ecology*, vol. 23, no. 1 (1998): 74–88.

several options. Rather, they are genetically encoded and respond to stimuli in the stepwise fashion of a computer program. Generation after generation of natural selection has continually fine-tuned the program, and a successful genetic program for nutritional and environmental requirements, host preferences, aquatic habitat, and mate selection criteria yields survival and progeny by enabling the female to undertake her tasks of feeding, mating, and laying eggs. Complementing this genetic programming is a vast array of sensors that she uses for monitoring her environment. For instance, mosquitoes can respond to remarkably subtle cues from chemical, temperature, air current, and light stimuli and can also detect ultraviolet, visible, and infrared components of the electromagnetic spectrum. The mosquito is complex, dynamic, successful, and, for its role in disease transmission, deadly.

General Biology of Insect-Borne Diseases

It is now well known by most people that mosquitoes transmit several pathogens—infectious organisms or agents—to humans. Ronald Ross, a British medical military doctor working in India, discovered the role of mosquitoes in malaria transmission in 1898.[43] Soon after Ross's discovery, one of the world's leading zoologists, Giovanni Batista Grassi, identified the genus of mosquito—*Anopheles*—that spreads the human disease.[44] Two or three years later, Major Walter Reed, an

[43] Dr. Ronald Ross was a somewhat reluctant hero of malaria control as he considered himself more of a poet and aesthete than a physician or public-health expert. As a British army doctor stationed in India where malaria was a significant problem for the military, he undertook numerous studies to understand how the disease was spread. Finally, in 1898, he discovered the mosquito's deadly role in malaria transmission.

[44] In fact, a long and acrimonious feud between Grassi and Ross ensued as to who had made the crucial discovery first. See Desowitz,

American Army medical doctor, discovered the role of *Ae. aegypti* mosquitoes in yellow fever transmission. These findings, which we now take for granted, were revolutionary and changed the whole approach to disease control.

Malaria and yellow fever are mosquito-borne diseases; however, mosquitoes are not the only insects that transmit disease to humans. Sand flies, for example, transmit leishmaniasis, a disease affecting hundreds of thousands of people in the Middle East, Africa, and Latin America.[45] Insect-borne diseases that affect humans fall into two general categories: anthroponoses and zoonoses. Anthroponoses are diseases in which the pathogens circulate or cycle between humans and the carrier insects. Other animals are generally not susceptible to these infections. Mosquito-borne anthroponoses include malaria, dengue, urban yellow fever, and bancroftian filariasis.[46] West Nile virus and leishmaniasis are zoonoses. The

---

R.S., *The malaria capers: tales of parasites and people* (New York: Norton, 1991), 192–98.

[45] Leishmaniasis is a parasitic disease caused by the parasitic protozoan *Leishmania*. Underreporting of the disease is widespread as it is largely a problem of rural remote areas. WHO estimates that around 2 million new cases of the disease occur each year, although only 600,000 are reported. Epidemics of the lethal visceral form of the disease have occurred in Africa, where mortality has been high. One study estimated that in epidemic-prone Sudan, 100,000 people died of visceral leishmaniasis out of a population of 300,000. WHO, "Leishmaniasis, background information," http://www.who.int/leishmaniasis/en/ (accessed Apr. 11, 2009).

[46] Bancroftian filariasis is probably better known to most readers as "elephantiasis." It is caused by parasitic worms that are transmitted by mosquitoes. When feeding on a human, the mosquito will inject the worm larvae into the human's blood system. The larvae will then travel to the lymph nodes in the body where they will mature. Repeated infections cause blockage of lymphyatics. This in turns causes gross swelling of the areas serviced by the lymphatic system. Most often this swelling occurs in the groin and legs. WHO, "Lymphatic filariasis,"

pathogens that cause them cycle between insects and animals; human involvement is often incidental. More often than not, zoonotic infections result from human intrusions into natural environments where the agents are cycling between animals and blood-feeding insects. The ecological gap separating humans from cycles of zoonotic infections can be bridged by human intrusions, but it also can be bridged in other ways. For example, a vector mosquito can bridge the gap. Two mosquito species that have invaded the United States within the last twenty to thirty years, *Ae. albopictus* and *Ae. japonicus*,[47] are rated along with one other native mosquito species, *Cx. pipiens*, as the most potent bridge vectors of West Nile virus in the United States. Bridging could occur when a bird brings the infection into an urban area where *Ae. albopictus*, which will bite both birds and mammals, bites the bird, becomes infected, then bites a human and transmits the infection.

The pathogens that cause insect-borne diseases are highly variable. Viruses cause dengue, yellow fever, and West Nile fever. Protozoan parasites cause malaria and leishmaniasis, and a helminth, a parasitic worm, causes filariasis. While the pathogens vary, the unifying characteristic of these diseases is that a vertebrate animal has the pathogen in its blood or tissues at the time an insect bites. And the vertebrate, of course, acquired the infection from an infected insect bite. This, in

---

http://www.who.int/mediacentre/factsheets/fs102/en/ (accessed Apr. 11, 2009).

[47] Turell, M.J. et. al., "An update on the potential of North American mosquitoes (Diptera: Culicidae) to transmit West Nile Virus," *J Med Entomol*, vol. 42, no. 1 (Jan. 2005): 57–62; Sardelis, M.R. and Turell, M.J., "*Ochlerotatus japonicus* in Frederick County, Maryland: discovery, distribution, and vector competence for West Nile Virus," *J Amer Mos Control Assoc*, vol. 17, no. 2 (Jun 2001): 137–41; Sardelis, M.R. et. al., "Vector competence of three North American strains of *Aedes albopictus* for West Nile virus," *J Amer Mos Control Assoc*, vol. 18, no. 4 (Dec. 2003): 284–89.

general, encompasses the cycle of insect-borne disease transmission.

The process of the insect acquiring the pathogen from a vertebrate, developing the pathogen within itself, and transmitting it to another vertebrate is referred to as biological transmission. As the insect imbibes the blood, the organism establishes itself in the insect and begins to develop. This developmental time within the insect, known as the extrinsic incubation period, requires several days, and the insect cannot transmit the pathogen until it is complete. After developmental change and/or multiplication within the insect, the pathogen moves to the insect's salivary glands or mouthparts. When the insect bites, it passes the pathogen to its vertebrate host. The insect bite does not result in immediate symptomatic disease. First, the pathogen must develop within the vertebrate, in the period of intrinsic incubation. Once intrinsic incubation is complete, the host will probably develop disease symptoms.

Of all the insect-borne diseases, those insect-borne anthroponoses that infect only humans have the greatest impact on human health because the pathogens of these diseases are the most highly adapted to humans. Dengue fever is an important anthroponosis, as is bancroftian filariasis, but year after year, malaria continues to be the most common and devastating of all insect-borne anthroponoses, infecting over half a billion and killing one to two million people per year.[48]

Zoonoses, though, can also exact a heavy toll on human populations. Leishmaniasis, for example, occurs in various species of animals, including humans. Human leishmanial infections are less common than many anthroponoses. Yet, because they occur in many animal species and can be transmitted by many species of sand fly, the leishmanias are widely distributed in the rural areas of many geographic regions. In some areas, leishmania-infected sand flies are abundant

---

[48] Snow, R.W., et al., "The global distribution of clinical episodes of falciparum malaria." *Nature*, Mar. 10, 2005, 214–17.

around human habitations. As the U.S. military discovered in Iraq, and before that in Afghanistan, where sand flies, humans, and the leishmanial agents occur together, leishmaniasis can be a serious human disease problem.[49]

For a disease to spread effectively within human populations, people must be susceptible to infection and, if infected, provide enough pathogens in circulating blood to infect insects when bitten. The insects must feed on humans, acquire infection from human blood, accommodate development of the organism within their bodies, and, at a later date, bite and infect another human.

Dangerous Associations of Insects and Humans

Just as certain pathogens infect only humans (the anthroponoses), some insects that transmit these pathogens prefer to feed on humans, a preference referred to as anthropophagy (human biting). Anthropophagy is a population phenomenon. It is a behavior that occurs so often across the whole population that it is characteristic of the species. However, anthropophagy is a much more complex phenomenon than just a selective preference for biting a human. When given the choice, many mosquito species that entomologists consider anthropophagic will choose to bite other mammals over humans.

Birds and mammals are mobile, and different species tend to be in specific environments at specific times of day or night. Adult mosquitoes are also highly mobile and actively move to particular environments at particular times. However, the mosquito's environment may not be that of the preferred host. It is important to understand that if, for a given species, there is

---

[49] In 2004, over 300 soldiers were infected with the leishmaniasis parasite, causing chronic, festering sores—now known as Baghdad boil—and it was expected to infect nearly 1,000 soldiers by the end of that year. Lockwood, D., "Soldiers Battle 'Baghdad Boil,'" *Nature Medicine*, vol. 10, no. 110 (2004).

divergence between host preference and host-seeking behavior, the species will likely fail to survive. For example, imagine that if, given the choice, a mosquito would prefer to bite waterfowl at night. Now at night, waterfowl rest on the surface of a pond. However, imagine that the innate behavior of the mosquito directs it not to hang out on the pond at night, but rather to go into the forest and to move upward into the tree canopy to seek its blood meal. This mosquito is not going to encounter any waterfowl wandering through the canopy; instead it will encounter roosting birds. Its innate behavior will always lead it away from its preferred host. It will never be near a waterfowl when it needs to feed and its chances of acquiring blood and producing eggs are lower than if its behavior had directed it to seek hosts on the pond. In essence, host preference and host-seeking behavior need to function in unison. Over evolutionary time, host-seeking behavior modulates biting preference and biting preference modulates behavior. Acquiring blood and producing viable eggs is the ultimate measure of success for the species.

For many mosquito species, there are broad groupings of hosts that a species will or will not bite. Yet, these are generalizations, and some mosquito species are so rigid in their preferences that they will hardly bite other than preferred hosts. While some species might opportunistically feed on some mammals even though they are primarily bird feeders, others might feed exclusively on birds. In the context of these behaviors, *Ae. albopictus*, feeding primarily on mammals (including humans), becomes a bridge vector of West Nile virus because about one in five of its blood meals are commonly from birds.

Regardless of the strategy the female mosquito uses to find a host, if it is generally successful, then her numbers will potentially increase. If the strategy is not successful, then her numbers will decline, perhaps to extinction. Rather like their numerous egg-laying strategies then, different species of mosquito have different strategies for acquiring blood, which are complex, flexible, genetically determined, and specific to each species. A strategy that takes a mosquito into the tree canopy at night to bite roosting birds will not result in this mosquito taking a meal of human blood inside a house. Likewise, a strategy of movement from a daytime resting site in the forest to an open area around a house, to the house, then indoors to bite will not result in a meal of bird blood in a tree canopy.

Within these broad groupings of hosts, some mosquitoes are very narrow in feeding and tend to bite with high frequency only one or two host species. The important malaria vector in Africa, *An. funestus*, is an example: it feeds almost exclusively on humans.[50] In almost every case of narrow host choices, the female mosquitoes move at the right time to be in presence of the host, and in that location, the mosquito's options for biting are largely limited to that host. Anthropophagy, then, doesn't mean that the mosquito bites humans because she finds them to be toothsome members of the animal kingdom; rather, she bites them because they are there when she needs to feed. She moves to the human habitation because of her innate responses to environmental cues, such as clearings around houses, human emanations, portals of entry, and resting sites within houses.[51] Furthermore, her movements occur according to a complex set of circadian (time-dependent) rhythms.

Most insects that transmit malaria, dengue, or filariasis tend to enter houses. These species can be characterized as anthropophagic—they bite humans—but they can also be characterized as endophagic—they bite indoors. These

[50] Mwangangi, J.M. et. al., "Blood-meal analysis for anopheline mosquitoes sampled along the Kenyan coast," *J Amer Mos Control Assoc*, vol. 19, no. 4 (2003): 371–75; Konate, L. et. al., "Zoophagia and alternative host selection of the malaria vectors in Senegal," *Parasite*, vol. 6, no. 3 (1999): 259–67.

[51] Costantini, C. and Diallo, M., "Preliminary lack of evidence for simian odour preferences of savanna populations of *Anopheles gambiae* and other malaria vectors," *Parassitol*, vol. 43, no. 4 (2001): 179–82.

relationships are exceedingly important because most malaria infections are acquired inside houses. Stopping mosquitoes from entering houses and biting indoors can prevent these cases.

Malaria: Parasites of Human Blood

The parasites that cause malaria come from the bite of an infected mosquito. In turn, the mosquito acquires malaria parasites from biting infected humans. Human malarias only occur in humans; no other warm-blooded animals are involved. The fact that only humans are involved has great significance for how the disease can be controlled.

Once in humans, the malaria parasites feed on blood. In the initial stages of infection, the parasites hide out and multiply in the liver. They then leave the liver cells and invade blood cells in repetitive cycles of reproduction and invasion of other blood cells. Symptoms and suffering of infected people accompany the repetitive cycles of parasite development in blood. The malaria patient begins experiencing cycles of uncontrollable shaking, chills, high fevers, and intense headaches, interspersed with hours of seeming normality. Within a red blood cell, the parasite devours cellular material as it multiplies. The infected cell is utterly destroyed when progeny rupture from it. Once free of the cell, each new parasite will then invade another cell. Cycle after cycle, day after day, the invasion of cells, consumption of cell contents, multiplication within cells, destruction of cells, and invasion of whole new populations of cells continues. The patient suffers more and more, growing ever more anemic (losing red blood cells), and the blood quickly loses oxygen-carrying capacity. Ultimately, the patient becomes exhausted and sick to the point of being comatose. At some point in this malaria infection, the process of organ and system failures can begin and death quickly follows.

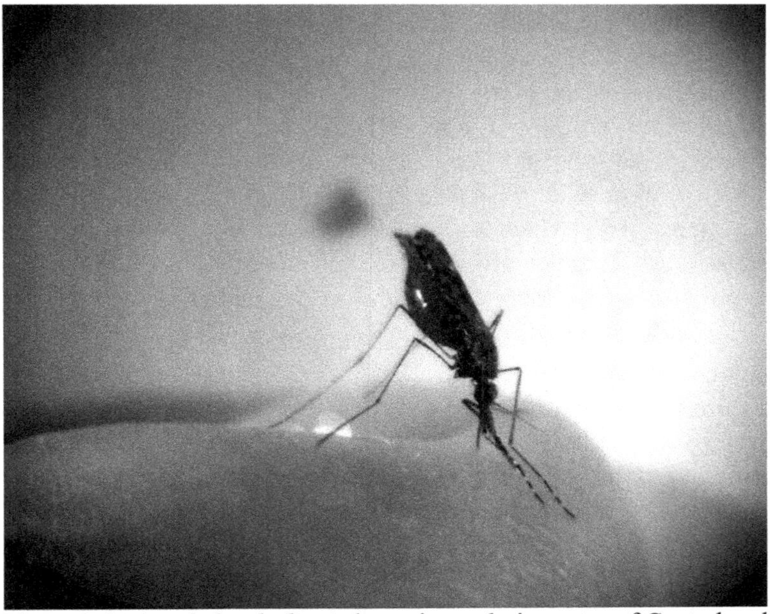

Figure 2-1: *Anopheles darlingi*, the major malaria vector of Central and South America, feeding on a human host.

The horror of infection is apparent in the enlarged spleen and liver that result from the body fighting to mount an immune counteroffensive and clean the blood supply. The liver filters the blood and becomes grossly enlarged with cellular debris. Antibodies are formed against the parasites, but there is no decisive win over this infection. Without treatment, the disease lingers, leaving the patient either dead or severely anemic. Even when fought to a standstill so the patient shows no apparent illness, the patient is still vulnerable to re-infection.

When the disease causes death, it is most frequently visited upon pregnant women and infants, whose deaths strike at the most cherished components of family life. By killing infants, the disease adds to high human fertility rates (defined as the ratio of live births to population, often expressed per thousand population per year) as parents plan larger families to compensate for children lost to malaria. These families may think, "we must do something about this disease before we lose more of our children," but they are not in a position to do much about it. Malaria is an ever-present threat, and it is entirely beyond the abilities of impoverished rural families to fight it. Members of this group are the most deeply affected by the disease because they have a poor understanding of it and few resources with which to protect themselves. Options for the heads of households are simple: have more children, or have no family to care for them when they are sick and/or elderly.

Malaria contributes to and sustains poverty. Frequent illness imposes costs from lost work and the expense of treatments. Continuous infections in adults leave breadwinners and caregivers sick and chronically anemic. Malaria infections in pregnancy stunt fetal development and impose a variety of decrements and medical conditions on the newborn. Malaria-induced anemia further impairs development and learning in children.

Malaria As Population Control

In the modern world of escalating fearful claims by practitioners of the environmental movement and pronouncements of ever-scarier scenarios, there are those who believe that human population growth poses a serious threat to life on earth. This fear was a basic building block of the modern environmental movement.

Paul Ehrlich reveals this fear in his attack on the use of DDT in malaria-control programs as the "export of death control" from rich countries to poorer countries that, in turn, allowed growth of those poorer populations.[52] Practitioners of such ideologies apparently believe that killing off large numbers of humans with natural infections like malaria and HIV/AIDS is helpful. However, this concept is not only immoral, it is also

---

[52] Ehrlich, *The Population Bomb*, 15.

incorrect. High malaria infection rates actually perpetuate high fertility rates, and high fertility rates propel the growth of human populations in malaria-endemic regions.

One outcome of the environmentalist agenda against insecticides that is little appreciated is the way in which it influenced a shift away from disease prevention and toward death prevention. In other words, vector control using DDT or other insecticides, which was the primary method of preventing disease, was increasingly abandoned in favor of treating the patient. This strategy ignores a fundamental relationship: without preventive measures, numbers of malaria cases increase, and, as a consequence, there are more and more cases to treat. With more and more cases being treated, there will be more and more selective pressure for drug resistance in malaria parasites. Alternatively, spraying houses works to directly reduce the number of malaria cases. Fewer cases allow increased survival of humans as disease and death decline. Fewer cases also reduces selective pressure for drug resistance in malaria parasites.

Initially, increased survival contributes to increased population growth, but control of disease also contributes to better health and greater economic productivity, which, in turn, can lead to improved standards of living and improved prospects for the future. In the long run, these trends, as judged by the history of countries that have eliminated malaria, can be expected to reduce human fertility rates. This is evidenced by trends of human population growth, economic development, diseases, and human fertility rates of countries with and without malaria. Improved economies are associated with reduced burdens of malaria. Likewise, growing wealth is associated with decreased human reproduction. Thus, if we look to what happened during the years of effective malaria prevention, we find that controlling malaria contributed strongly to healthier families, healthier economies, and reduced human fertility rates. Ehrlich could have looked at the history of insect-borne diseases such as malaria and yellow fever in his native United States to understand these relationships. Sustaining such progress, however, depends on good governance and public institutions such as the rule of law.

Malaria Control in the United States

The politics of disease control has a long and convoluted history in the United States and other countries of the world. Many of the great diseases, such as malaria, yellow fever, dengue, typhus fever, and plague, were endemic to the United States. Indeed, arthropod-borne diseases ravaged many parts of the United States, claiming thousands of lives and harming economic progress. We review this history of disease control in appendix 1.

In the late nineteenth century, even before discovering how some diseases were being transmitted, the diseases themselves were declining in many rapidly developing countries. Increased development of agricultural land reduced the abundance of larval habitats of mosquitoes, and better housing reduced mosquitoes' access to people inside houses. Also, drugs for treating diseases, such as quinine, became more available. For these and similar reasons, malaria was declining in the rapidly developing countries.

DDT was the final step in malaria eradication in both the United States and Europe. After World War II, the Rockefeller Foundation, the Tennessee Valley Authority, and the United States Public Health Service immediately added DDT into their mosquito-control efforts in the United States. Similar programs with broad international support were initiated in Europe.

While today some commentators have argued that even without DDT mosquito control, efforts in the United States would have succeeded,[53] the literature from the time does not

[53] This argument states that malaria would have been eradicated through the lifestyle changes that had occurred in the United States,

show such optimism. The 1940s mosquito-control workers thought they had achieved what they could with the tools they had, and the appearance of a safe, effective, residual pesticide was a godsend. As explained in a U.S. public-health document for the period of 1944–45,

> Drainage and larviciding are the methods of choice in towns of 2,500 or more people. But malaria is a rural disease. Heretofore there has been no economically feasible method of carrying malaria control to the individual tenant farmer or share cropper. Now, for the first time, a method is available—the application of DDT residual spray to walls and ceilings of homes.[54]

The mosquito-control officers used DDT in two ways—as a residual insecticide on the walls of houses and as a larvicide.[55] The results were dramatic: in 1952, malaria was officially pronounced eradicated from the United States.

---

including the environmental changes wrought by the Tennessee Valley Authority and the TVA's dedication to designing water projects that limited mosquito breeding through the application of larvicides, such as Paris Green. See Humphreys, M., *Malaria: poverty, race and public health in the United States* (Baltimore: Johns Hopkins University Press, 2001), 2 ; Spielman and D'Antonio, *Mosquito*, 151–54. This is not to say that pesticides weren't used, even unofficially. Spielman relates that when he worked for the TVA in the 1950s, he investigated a sudden decline in malaria incidence in the Black Swamp of Northern Alabama. It turned out that the sharecroppers were "borrowing" toxaphene (a far more toxic substance than DDT) from the cotton farms and dusting their houses with it.

[54] U.S. Public Health Service, "Malaria control in war areas, 1944–1945," Federal Security Agency, U.S. Public Health Service, 134.

[55] Ibid., 17–18. Briefly, though the program was slated to spray 1,222,225 homes by the end of the fiscal year, only 300,000 were sprayed, and about 8 oz. of DDT was used per house. A house was estimated to have an average size of 2,200 sq. ft. and DDT was sprayed at 100 mgs. per sq. ft.)

The Magical White Powder in Poor Countries

The use of DDT for controlling malaria evolved rapidly after its broad public-health use in WWII. National programs for controlling malaria in endemic countries used DDT from the mid-1940s to the late 1950s. In 1955, the eighth meeting of the WHA directed WHO to coordinate a global malaria-eradication program that was largely based on the use of DDT for spraying houses. This program got fully underway over the next three to four years. A developed global eradication program existed from about 1959 to the end of 1969.

Historical data from many countries show the huge impact of spraying houses with DDT. We will mention only a few. Examples of country data reviewed here are not exceptional. Rather, we have selected examples that are representative of what occurred in many countries around the world.

The vector-control programs in southern Africa quickly adopted DDT after the end of WWII, and as a direct consequence, malaria cases declined rapidly. According to Mabaso et al., in the Transvaal Province of South Africa,[56] hospital admissions "fell from 1,177 cases during the 1945–46 transmission season to 601 in 1946–47 coinciding with the availability of DDT in 1946, and falling to 454 in 1948 and to a low of 61 cases in 1951."[57] Similar successes were echoed in other southern African countries wherever DDT was used.

In the Americas, the period from 1943 to 1945 in Guyana was characterized by over 37 percent and 27 percent of rural and

[56] After South Africa's political transformation to democracy, provincial borders were redrawn and the Transvaal Province was turned into four new provinces—Gauteng Province, Mpumalanga Province, Limpopo Province, and North West Province. Of these only Mpumalanga Province and Limpopo Province are now considered malarial.

[57] Mabaso, M. et. al., "Historical review of malarial control policies in southern Africa with emphasis on the use of indoor residual house-spraying," *Trop Med Int Hlth*, vol. 9, no. 8 (Aug. 2004): 852.

urban people, respectively, having malaria.[58] Along with this high infection rate, the death rate of newborns was 126 per 1,000 live births. By contrast the death rate of newborns in the United States in 2004 was 4 per 1,000 live births.[59] Guyana started spraying DDT inside houses in mid-1946. The stunning impact of DDT was revealed in disease and death statistics for the period from 1948 to 1949. As a result of house spraying alone, infant mortality declined by 39 percent. In urban areas, malaria infections declined by 99 percent. Even in the highly endemic rural areas, malaria infections declined by 96 percent. These statistics are all the more telling because the distribution of malaria drugs was not part of the early malaria-control program in Guyana.

In 1945, the government of Bolivia started using DDT against *Ae. aegypti*, the mosquito vector of the dengue and yellow-fever viruses.[60] By 1947, Bolivia had eradicated this mosquito. Bolivia's quick success stimulated the Pan American Health Organization to begin a hemisphere-wide program to eradicate *Ae. aegypti*. By the early 1950s, many countries had eradicated or greatly reduced the distribution of this dangerous mosquito. Through their successes, the risks of dengue and yellow-fever epidemics largely disappeared from Central and South America.

The early tests of DDT for malaria control were spectacular. It seemed that DDT might serve to free humanity of an enormous disease burden. Wherever DDT was sprayed on house

---

[58] Curtis, C. (Ed.), *Demarara doctor: An early success against malaria. The autobiography of a self-taught physician, George Giglioli, 1897-1975.* Smith-Gordon and Company, Limited, London, UK. 270p.

[59] World Health Organization, "World Health Statistics 2009," WHO, Geneva, 2009, http://www.who.int/whosis/whostat/2009/en/index.html (accessed Sept. 24, 2009).

[60] Severo, O., "Eradication of the *Aedes aegypti* mosquito from the Americas," in *Yellow fever: A symposium in commemoration of Carlos Juan Finlay* (Philadelphia: The Jefferson Medical College of Philadelphia, 1955), 39–58.

walls, malaria rates dropped. Although there were critics of the vector-control programs,[61] the world had never seen a chemical with such a powerful impact on malaria, and DDT proved equally effective against other insect-borne diseases, e.g., dengue and leishmaniasis.

The number of malaria cases on the Indian subcontinent was far higher than in South Africa, and the scale of success in malaria control was far more dramatic. Before DDT was introduced for malaria control in India there were as many as 75 million malaria cases per year.[62] In 1951, India's malaria-control program began to use DDT and soon after saw some spectacular health benefits. Malaria cases were reduced to 286,962 by 1969.[63] Fifty-three years after Ross made his ground-breaking discovery in India that mosquitoes transmit the malaria parasite, the country was finally able to use his theory and methods with success thanks to DDT.

---

[61] Some critics warned that the programs of DDT spraying and effective malaria control were advancing the American ideals of capitalist development and undermining communism. For example, Litsios notes that "the real (or imagined) fear that the Italian government would be won over by the communists at the next election was used to justify continued external funding for malaria control, even though this was no longer technically required." Speaking in 1948 at the third joint session of WHO/FAO, the Italian malaria expert, Missiroli, noted that "Africa cannot be fully exploited because of the danger of flies and mosquitoes; if we can control them, the prosperity of Europe will be enhanced." In spite of these criticisms and the concerns over imperialism, DDT was still saving thousands, if not millions of lives. Litsios, S. "Malaria control, the cold war and the post war reorganization of international assistance." *Med Anthropol*,17, 3:255-278.

[62] Lals, S., Sonal, G.S., Phukan, P.K., "Status of malaria in India." *J Indian Acad Clinical Med* 5 (2000):19-23.

[63] WHO, Executive Board, "The place of DDT in operations against malaria and other vector-borne Diseases," 42nd Session, Appendix 14, (Geneva, 1971), 178.

After DDT was introduced to malaria control in Sri Lanka (then Ceylon), the number of malaria cases fell from 2.8 million in 1946 to just 110 in 1961, and deaths from malaria dropped from 12,587 to zero during this same period.[64] Similar spectacular decreases in malaria cases and deaths were seen in all the regions that began to use DDT. The newly formed Republic of China (Taiwan) also adopted DDT use in malaria control shortly after WWII. In 1945 there were over 1 million cases of malaria on the island, but by 1969 there were only nine cases. Shortly thereafter, the disease was eradicated from the island (and it remains eradicated).[65]

Countries that started their own programs and pioneered house-spray methods laid the groundwork for what became the WHO-coordinated global malaria-eradication program. Over a third of a billion people were freed from endemic malaria even before the WHO-led effort began. After global eradication was formally under way in 1959, another two-thirds of a billion people were freed from endemic malaria over the next nine years.

Nepal was just one of several countries to benefit from the global malaria-eradication program. Technical advisors on behalf of the United States Overseas Mission, which later became USAID, arrived in 1959 to start the malaria project. By 1960, geographic reconnaissance (mapping of houses), training, and organizing were complete. In 1960, spraying of houses in the central zone began. In 1962, preparatory work started in the eastern zone, and spraying began in 1964. Preparatory work started in the western zone in 1963, and spraying started there in 1965.

In the 1950s, malaria was Nepal's most serious public-health problem, and the burden of deaths fell most heavily on children. In the early 1950s there were more than two million cases of malaria in Nepal each year with a mortality rate of ten percent.

---

[64] Ibid.

[65] Ibid.

Malaria contributed to increased deaths from other diseases as well. Some have opined that "dramatic, nationwide reduction of malaria [in Nepal] was perhaps the greatest technical and logistic triumph of the 1960s."[66] That triumph over malaria was based entirely on the use of DDT. From more than two million cases a year, only 2,468 cases were found in 1968. Before the malaria-control program, life expectancy in Nepal was 28 years. By 1962, life expectancy was 33 years, and by 1970, it was 42.3 years.[67]

DDT miraculously changed the global distribution of endemic malaria. The total cost of the multi-year program in Nepal, which oversaw vast improvements in health, dramatically reduced death, increased longevity, and ultimately, vastly increased agricultural productivity and wealth, was $13 million (around $76 million in today's currency).[68] By the time the global eradication program ended in 1969, almost one billion people had been freed of endemic malaria by DDT-sprayed house walls. This achievement is unequaled in the history of arthropod-borne disease control. In those countries that coupled malaria eradication with economic growth and development, such as the United States, the disease remains banished.

---

[66] Skerry, C.A., Moran, K., and Calavan, K.M., *Four decades of development: The history of U.S. assistance to Nepal 1951–1991*, (Kathmandu, Nepal: USAID, 1991), 141.

[67] Ibid, 74.

[68] Ibid, 50–53 and 141–150. Malaria was an enormous problem in many areas, and this problem stopped people from occupying land and producing crops. With control of disease, previously unoccupied areas suddenly became available for agricultural productivity. As land became available, there were large movements of populations into new areas.

Summary

The discovery that *Anopheles* mosquitoes transmit malaria gave public-health workers new options to control the disease by creating a chemical barrier between man and mosquito. But the knowledge that *Anopheles* mosquitoes transmit malaria was not sufficient. Malaria-control scientists had to know how the mosquito behaved—where and when it fed, rested, and laid eggs. In particular, the mosquito's house-entering behavior would prove to be of critical importance in the use of indoor spraying for preventing disease transmission. In addition, in order to control malaria, scientists needed to understand the parasite and the medicines to which it was susceptible.

The introduction of DDT to malaria control was so effective that it eradicated the disease in Europe and the United States and in some island states, such as Taiwan. The eradication program saved millions of lives and lifted the threat of this ages-old disease from many hundreds of millions more. In countries that benefited from economic growth and development, the eradication lasted. Less-developed countries with poor medical infrastructure and an inability to sustain their house-spraying programs because of continual pressures from environmentalists and international donors meant that slowly but surely, the disease crept back in those areas.

As we explain in the next chapter, however, while DDT was highly successful, there were widespread misunderstandings about how the chemical actually worked to control the disease. The persistence of these misunderstandings has compromised malaria-control programs.

Chapter 3

Misunderstanding How DDT Works to Control Malaria

DDT works in unexpected ways, and incorrect assumptions about its mode of action have been costly. We have therefore compiled a detailed account of the scientific history and evidence of how DDT works. Readers who find these details tedious may chose to simply read the overview below, which provides in a few paragraphs the nub of what we explain in the next few pages. The table given in this chapter will also provide the reader with a useful summary of some of the key studies and publications that describe DDT's unique modes of action.

Overview

Control of malaria by indoor residual spraying (IRS) of DDT on house walls has been the topic of extensive studies around the world. Such studies began in the early 1940s, even before DDT was broadly used against malaria. The results of such studies, covering decades of malaria control efforts, are described and presented in considerable detail in this chapter. Early descriptions of DDT as an almost miraculous insecticide that killed insects and continued killing for months after it was sprayed set the stage for huge misunderstandings about how it actually functioned to prevent transmission of malaria inside houses. This misunderstanding became deeply embedded in malaria control lore as a result of a mathematical model developed by Sir George Macdonald. Macdonald's model was published over a period of years in the late 1940s and early 1950s. The model was based on the erroneous assumption that DDT killed mosquitoes and reduced mosquito longevity below critical thresholds needed to maintain disease transmission.

In contrast to Macdonald's assumption of DDT killing mosquitoes, the literature is filled with study after study, in country after country, showing DDT functioned more as a repellent. It prevented malaria, not by killing mosquitoes, but by stopping mosquitoes from entering houses and biting people while they slept. As a secondary action, if the mosquito was not repelled and entered the sprayed house, then DDT's contact irritant action often caused them to leave the house immediately. Field studies have shown that DDT's repellent action may stop as much as 95 percent or more of mosquitoes from entering a sprayed house. Furthermore, computer modeling suggests that non contact repellent and contact irritant actions would each need to function below 30 percent level of effectiveness before contact toxic actions could become the dominant mode of DDT action against malaria mosquitoes. This prioritization of actions is largely based on what actually occurs when mosquitoes approach a sprayed surface. For non contact repellent action, not even fleeting contact with DDT is required before the mosquito begins moving away from a treated surface. For contact irritant action, brief physical contact with DDT will stimulate the mosquito to respond by moving away from the treated surface. Often, the irritant response is so rapid as to preclude any toxic action. On the other hand, taking up a toxic dose through physical contact with a treated surface only occurs after more prolonged physical contact.

The appearance of DDT resistance in malaria mosquitoes contributed to the persistent misunderstanding (described above) of how DDT worked. Many thought that if DDT did not kill mosquitoes then there would be no selection for DDT resistance. This was, of course, faulty logic. Even in the 1950s there was clear evidence that mosquito resistance was evolving from agricultural use of DDT, not DDT sprayed on house walls. Agricultural use of DDT selected for resistance in mosquitoes because the larvae were present in bodies of water that became contaminated when DDT was sprayed on crops. The repellent and irritant actions of DDT against mosquitoes were useless under those circumstances because mosquitoes could not avoid contaminated waters. Mosquitoes either became resistant or died. Ultimately, resistant mosquitoes sprang up in many areas where DDT was used on crops, but not if DDT was used only for malaria control. Regardless, and in line with DDT's multiple modes of chemical action, DDT continued to provide high levels of malaria control in some localities in spite of the malaria mosquitoes being DDT resistant.

The way DDT actually works to prevent malaria transmission inside houses is important. The dominance of repellent and irritant actions should be considered in decisions to replace DDT in IRS programs where mosquitoes are DDT resistant. Clearly such decisions should be based on whether DDT still controls malaria through repellent and irritant actions, not on whether it kills mosquitoes. DDT's multiple chemical modes of action are also important in the selection of test methods for screening new chemicals to find a legitimate DDT replacement. To this day, WHO does not consider IRS with DDT to accomplish anything but kill mosquitoes. This uncompromising and exclusionary policy is fully documented in WHO technical documents and in WHO methods for evaluating insecticides for IRS. Under this policy, a chemical repellent action is considered undesirable. There is now a very great need to fully recognize that non-contact repellent, contact irritant, and contact toxic agents each make important contributions to the control of malaria. WHO's exclusionary policy that the only purpose for spraying insecticides on house walls is to kill mosquitoes has harmed public health and it must be changed.

DDT Controversies

The chemical action of DDT was debated from the very beginning of DDT-spray programs. Since the mid-1940s, leading malaria scientists observed that DDT had powerful actions on mosquito behaviors. But their observations stood silently in the literature, ignored by public-health administrators and science bureaucrats. Today, there is an even greater body of evidence, which is consistent in countries around the globe and across decades of field and laboratory studies. The evidence shows that DDT is a unique compound with unique actions. It repels mosquitoes and stops them from entering houses. Yet, that evidence has been ignored, and the belief that DDT controls malaria only by killing mosquitoes continues to dominate international policies for malaria control. This is a misunderstanding of huge proportions.

The misunderstanding occurred because DDT does, in fact, kill mosquitoes. Another reason for the misunderstanding was the wide acceptance of an epidemiological model for malaria that attributed the success of spraying houses to its killing of mosquitoes. Given the model's assumption, it was quite reasonable to assume that DDT resistance would neutralize the chemical's value for disease control. Thus, the malaria-eradication program was based on a need for rapid eradication before mosquitoes became DDT resistant. Years later, the fear of a growing resistance of mosquitoes to DDT was put forward as one of several reasons for ending the eradication program.

The Role of a Mathematical Model and DDT Resistance

Efforts to mathematically model the interactions of malaria and DDT accompanied the enthusiasm of malaria workers for using DDT. Sir George Macdonald, a talented mathematician who worked at the Ross Institute in Sri Lanka (then Ceylon), developed an elegant mathematical model to describe major components of malaria transmission and malaria epidemiology. His model treated DDT as an insecticide that functioned by killing mosquitoes. In his model, Macdonald allowed that malaria mosquitoes would die once they entered houses and came to rest on DDT-sprayed surfaces. He further allowed that the killing action was sufficient to lower the overall population longevity of malaria mosquitoes to such an extent that, on average, the females did not live long enough to complete development of malaria parasites. This meant that with DDT, the mosquitoes would no longer become infectious and malaria transmission would stop—all as a result of killing mosquitoes.[69] Then, in 1951, malaria-control workers discovered a population of mosquitoes that would not die from DDT at concentrations that should have been lethal. This finding raised the specter of a time when public-health officials would no longer be able to use DDT to control malaria.

The relationship between the agricultural use of DDT and DDT resistance in a malaria vector was apparent in even the first reported case of resistance (see footnote 72 referring to Laird et al below). Between the 1940s and 1970s, hundred of millions of pounds of DDT were used in agriculture, with peak use occurring in 1959, while public-health programs used only small quantities inside houses. The amount of DDT produced in the United States alone (roughly 85,000 tons in 1962, year of peak production) greatly exceeded the peak amount used in the whole world for spraying houses.[70] Although many claimed that use of DDT in malaria-control programs was the cause of resistance, a WHO assessment placed the blame for DDT resistance in mosquitoes squarely on use of DDT in agriculture. DDT resistance continued to evolve in different mosquito species in various localities. However, WHO properly opined that "resistance had been mainly induced by the contamination of the larval breeding areas with insecticides applied to agricultural crops."[71]

---

[69] Macdonald, G. and Davidson, G. "Dose and cycle of insecticide application in the control of malaria," *Bull Wld Hlth Org*, vol. 9 (1953): 785–812.

[70] Agency for Toxic Substances and Disease Registry, "Toxicological Profile for DDT, DDE and DDD," U.S. Dept. of Health and Human Services, Washington, DC, 221, http://www.atsdr.cdc.gov/toxprofiles/tp35-c5.pdf (accessed Apr. 12, 2009).

[71] Brown, A.W.A., et. al., "Review Article. Malaria eradication and control from a global standpoint," *J Med Entomol*, vol. 13, no. 1 (1976): 1–25.

The 1951 report of resistance in a malaria mosquito, *An. sacharovi*,[72] not only sent an alarm through the malaria-control community, it also launched the global eradication program. The appearance of DDT resistance stimulated many to think that it was necessary to use DDT quickly and eradicate malaria before DDT resistance rendered DDT useless. The Eighth World Health Assembly in 1955 instructed WHO to coordinate a program for global eradication of malaria. Technical guidelines for eradication were accepted in 1956.[73] The conversion of existing programs into eradication efforts was completed by 1958, and eradication programs were officially underway in 1959.

Misunderstanding DDT Effectiveness

In the conduct of the global malaria-eradication program, WHO based all its major initiatives on a fundamentally erroneous assumption. Though elegant and accurate in so many ways, Macdonald's model was wrong in its fundamental assumption that DDT functions by killing mosquitoes. This error is all the more unfortunate because Macdonald himself doubted the accuracy of his core assumption about DDT's mode of action.[74] Yet, his model was embraced so completely that Macdonald's own concerns were swept away.

---

72 "Resistance to the DDT deposits sprayed on house walls for malaria control was first discovered in 1951, in *Anopheles sacharovi* at Nauplion in the Peloponnese peninsula of Greece, a locality where DDT had been applied to rice fields since 1946." Laird, M. and Miles, J.W., eds, *Integrated mosquito control methodologies, vol. 1* (New York: Academic Press, 1983), 188.

73 WHO, "Sixth Report, Expert Committee on Malaria," Tech. Rep Series, 1957, No. 123, pages 19, 30–33.

74 Macdonald and Davidson, "Dose and cycle of insecticide application," 786. The authors state in the introduction of their 1953 paper that most of their evaluations of DDT and other insecticides were

The model was based on a simple and easily understood assumption: DDT killed mosquitoes. The fact that some mosquitoes became resistant to DDT seemed to confirm that DDT selectively killed mosquitoes—thereby selecting for resistance. Yet, this resistance was considered to be mostly as a result of DDT use in agriculture, where mosquitoes were much more likely to be exposed to lethal and sublethal doses of DDT. Furthermore, in areas such as the Amazon Basin, where DDT was seldom used in agriculture, continuous use of DDT for malaria control in large geographical areas against different mosquito species did not select for resistance.

We might also conclude that Macdonald's explanation was widely accepted because most malaria workers didn't understand the mathematical model, so acceptance was easier than arguing about the weakness of the underlying assumptions. This broad acceptance reveals a serious and ongoing problem with the role of models. In the case of DDT and malaria, Macdonald's model, reinforced by the appearance of DDT resistance, was assumed an axiom of malaria eradication. But it was really more of a belief, since it was not the result of controlled experimentation and analysis, but merely a convenient assumption. As a belief, it rallied adherents to its defense. Opposing views of how DDT functioned were ignored or actively suppressed.

Some might argue that it is preposterous to claim that a major global program could have been based on a misunderstanding. One could argue that the whole scientific

---

based on preliminary anopheline observations followed by observation of reduction of malaria. They thought more precise data would require much expanded experimentation, which was not practicable. They ended their introduction with the following statement: "Tentative conclusions are presented, but the one most clearly brought to light is that a considerable expansion of our knowledge is urgently needed—an expansion in the fields of basic theory, anopheline habits, the physical chemistry of insecticides, and their modes of action." This is clear statement of doubt about their own conclusions and analyses.

community could not possibly have erred on such a fundamental issue as DDT's mode of action. In fact, at the very beginning of house-spraying programs, many scientists did correctly identify DDT's primary modes of action as a repellent and irritant.[75] Furthermore, through all the decades of DDT use, fieldwork continued to show that DDT's suppression of malaria was not by killing mosquitoes. The problem was not that scientists didn't correctly identify how DDT functioned; the problem was that, at best, other scientists, bureaucrats, and public-health officials ignored the work and publications of those scientists. Perhaps the answer for why this occurred is best explained by the old maxim, "it's hard to argue with success." Regardless of how DDT actually worked, when sprayed on house walls, its control over malaria was spectacular.

Early Scientific Reports of DDT's Effect on Mosquitoes

Researchers studied the actions of DDT on malaria in the United States as early as 1943.[76]

---

[75] In the next section we will cover in considerable detail some of those initial studies and list others that showed how DDT exerted more powerful repellent and irritant than toxic actions. One of the first of those studies was one conducted by Metcalf and coauthors (Metcalf, R.L. et. al., "Observations on the use of DDT for the control of *Anopheles quadrimaculatus*," *Pub Health Rep*, vol. 60, no. 27 [July 1945]: 753–774), but there were many others. Later, we will list a number of studies from different countries over many years that continued to show that DDT was functioning mostly as a spatial repellent and as a contact irritant.

[76] Metcalf et. al., "Observations on the use of DDT."

DDT modes of action: Selected studies

Researchers	Study Dates	Study Location	Findings
Metcalf, R.L., et al.[77]	1943–1944	Tennessee Valley, U.S.	Differences in numbers of mosquitoes resting in DDT-treated barns, houses, and nail kegs outside showed mosquitoes were not entering DDT-sprayed structures and not resting on sprayed surfaces. These behavioral responses typify the repellent and irritant actions of DDT residues.
Gahan, J.B. & Lindquist, A.W.[78]	1944	Arkansas, U.S.	Unsprayed nail kegs, which served as resting sites for mosquitoes, were placed around DDT-treated buildings. While no mosquitoes were found in DDT-treated buildings, populations in nail kegs were unchanged, suggesting that DDT had powerful repellent properties that kept mosquitoes out of sprayed buildings but did not kill them or reduce overall mosquito-population numbers.

[77] Metcalf, R.L., et al., "Observations on the use of DDT for the control of *Anopheles quadrimaculatus*," *Pub Hlth Rep*, vol. 60, no. 27 (July 1945): 753–74.

Researchers	Study Dates	Study Location	Findings
Tarzwell, C.M. & Fisk, F.W.[79]	1944–1947	Georgia, U.S.	Demonstrated that DDT would kill mosquitoes after prolonged contact, that the insecticide remained active for over eleven months, and that there was a "reduction of the number of resting mosquitoes in buildings." Authors opined that repellent action keeping mosquitoes out of buildings "might well be one reason for the protection afforded by DDT residual sprays."
CDC Bull.[80]	1950	U.S.	Surveys were conducted in thirteen southeastern states, and 65,000 houses were inspected. DDT effectiveness was evaluated based on inspections of randomly selected sprayed and unsprayed houses for the presence of malaria mosquitoes. Sprayed houses were mostly free of mosquitoes for an estimated reduction of 90 percent in sprayed vs. unsprayed houses.

[78] Gahan, J.B. and Lindquist, A.W., "DDT residual sprays applied in buildings to control *Anopheles quadrimaculatus*," *J Econ Entomol*, vol. 38, no. 2 (1945): 223–30.

[79] Tarzwell, C.M. and Fisk, F.W., "Observations on the nighttime resting and biting habits of anopheline mosquitoes in DDT-treated and -untreated buildings," *Publ Hlth Rep*, vol. 62 (1947): 84–94.

Researchers	Study Dates	Study Location	Findings
Kennedy, J.S.[81]	1947	Wiltshire, U.K.	Laboratory studies involving the treatment of chambers with DDT proved DDT works "in two contradictory ways simultaneously . . . as a lethal agent . . . and as an excitant and thereby sometimes a repellent." "It appears that the success of residual DDT may be due partly to destruction of mosquitoes but partly also to their being repelled from the vicinity of the host."
Gebert, S.[82]	1948	Mauritius	Observed repellent effect of DDT that stops *An. funestus* from entering houses.
Fay, R. & Sheppard, E.[83]	1949	Georgia, U.S.	Confirmed Kennedy's findings and demonstrated that mosquitoes must be in contact with DDT for more than 20 minutes for more than 50 percent mortality.

---

80 CDC Bulletin, U.S. Public Health Service, Communicable Disease Center, Atlanta, Georgia, Jan. 1950, p. 11.

81 Kennedy, J., "The excitant and repellent effects of mosquitoes of sub-lethal contacts with DDT," *Bull Entomol Res*, vol. 37 (1947): 593–607.

82 Gebert, S., "Notes on certain aspects of the action of DDT residual sprays, and on the partial treatment of dwellings as a means of anti-anopheline protection," *Trans Roy Soc Trop Med Hyg*, vol. 42, no. 3 (Nov. 1948): 295–97.

Researchers	Study Dates	Study Location	Findings
Gabaldon, A.[84]	1949	Venezuela	Field studies demonstrated that DDT spraying greatly reduces malaria and that "as soon as DDT is sprayed anophelines disappear from houses. This may be the consequence of what seems to be its repellent action." Gabaldon was the first to identify contact irritant and non-contact repellent actions as separate chemical actions of DDT residues.
Muirhead-Thompson, R.C.[85]	1951	Data from many studies in different countries	Reported on many studies observing that DDT on house walls exerts strong repellent and irritant actions on malaria mosquitoes.

[83] Fay, R. and Sheppard, E., "*Anopheles quadrimaculatus* activity patterns in the laboratory on untreated and DDT-treated surfaces," *J Nat Mal Soc*, vol. 9, no. 2 (1949): 147–58.

[84] Gabaldon, A., "The Nation-wide campaign against malaria in Venezuela. Part I," *Trans Roy Soc Trop Med Hyg*, vol. 43, no. 2 (1949): 113–32; Gabaldon, A., "The Nation-wide campaign against malaria in Venezuela. Part II," *Trans Roy Soc Trop Med Hyg*, vol. 43, no. 2 (1949): 133–64; Gabaldon, A., "The effect of DDT on the population of anopheline vectors in Venezuela," *Riv Parassitol*, vol. XIII, no. 1 (1952): 24–41.

[85] Muirhead-Thomson, R.C., *Mosquito behaviour in relation to malaria transmission and control in the tropics* (London:, Edward Arnold & Co., 1951) 219 pp.

Researchers	**Study Dates**	**Study Location**	**Findings**
Van Thiel, P.H.[86]	1951	Indonesia	In experimental huts, on average 40 mosquitoes were found per day; however, after spraying, hardly any were found, dead or alive, indicating strong spatial-repellent actions that keep mosquitoes out of houses.
Macdonald, G. & Davidson, G.[87]	Published 1953	Data from many countries	Developed a mathematical model that was widely adopted as a principle for DDT use in malaria control. Macdonald & Davidson's model assumed DDT functions by killing mosquitoes and proposed increasing the dosage of DDT so that a lethal dose is absorbed before the mosquito is repelled.
Coluzzi, A. & Coluzzi, M.[88]	1961	Italy	The authors described the irritability effects of DDT observed in *Anopheles* mosquitoes.

[86] Van Thiel, P., "The repellent effect of insecticides, especially of contact insecticides," *Doc Neerl Indones Morb Trop*, vol. 3 (1951): 117–25.

[87] Macdonald, G. and Davidson, G., "Dose and cycle of insecticide application in the control of malaria," *Bull Wld Hlth Org*, vol. 9 (1953): 785–812.

[88] Coluzzi, A. and Coluzzi, M., "Sull'irritabilita degli *Anopheles* al DDT in rapporto all'eradicazione della malaria," *Riv Malariol*, vol. 40 (1961): 35–40.

Researchers	**Study Dates**	**Study Location**	**Findings**
Zulueta, J.[89]	1959, 1960, 1961	Greece, Italy, Romania (1959)	In a review of several papers, de Zulueta described how DDT interrupts transmission of malaria by its irritant actions, i.e., it acts to "drive most of them away." De Zulueta stated, "Against the general judgement, that the irritant effect of DDT is a disadvantage in the use of this insecticide, the author believes, with others, that irritability" reduces mosquito contact with man and achieves "as a consequence, the interruption of transmission."
Hecht, O. & Hernandez, J.[90]	1960	Mexico	Documented malaria mosquitoes avoided DDT residues. Observations relate to spatial repellent and contact irritant actions.
Vigueras, O.M. & Corzo, J.H.[91]	1960	Mexico	Observed behavioral avoidance of DDT residues by malaria mosquitoes. In other words, DDT functions as a spatial repellent and contact irritant.

---

[89] Zulueta, J., "The irritability of mosquitoes to DDT and its importance in malaria eradication," *Riv Malariol*, vol. XLI (1962): 169–78.

[90] Hecht, O. and Hernandez, J., *CNEP Boletin*, vol. 4, no. 2 (1960): 93–106.

Researchers	Study Dates	Study Location	Findings
Zulueta, J.[92]	1961, 1963	Uganda	Presented data from experimental hut studies in Uganda showing DDT interrupts malaria transmission through a spatial repellent action that stops vectors from entering sprayed houses. Clearly discriminated between contact-irritant and non-contact repellent actions of DDT. De Zulueta stated that the repellent action has "great operational significance."
Duret, J.P.[93]	1964	Panama	Observed behavioral avoidance of DDT residues by malaria mosquitoes. In other words, DDT functions as a spatial repellent and contact irritant.
Slooff, R.[94]	1964	New Guinea	Observed fifty percent fewer malaria vectors in houses sprayed with DDT compared with unsprayed houses.

---

[91] Vigueras, O.M. and Corzo, J.H., 1960. *CNEP Boletin*, vol. 4, no. 1 (1960): 82–85.

[92] Zulueta, J., et. al., "The results of the first year of a malaria eradication pilot project in northern Kigezi (Uganda)," *E Afr Med J*, vol. 38, no. 1 (1961): 1–26; Zulueta, J. and Cullen, J.R., "Deterrent effect of insecticides on malaria vectors," *Nature*, vol. 200, no. 4909 (1963): 860–61.

[93] Duret, J.P., *Boletin Of Sanit Panamer* Septiembre 1964: 250–67.

Researchers	Study Dates	Study Location	Findings
Shalaby, A.M.[95]	1966	India	Using innovative approaches to hut studies, Shalaby showed that DDT functions as a powerful spatial repellent and contact irritant.
Smith, A. & Webley, D.J.[96]	1969	Taveta, Kenya	Presented data confirming de Zulueta's 1963 study showing DDT functions as a repellent, an irritant, and a toxic agent.
Ismail, I.A.H., et al.[97]	1975	Thailand	Documented malaria mosquitoes avoided entering DDT-sprayed houses.
Nutsathapana, S., et al.[98]	1986	Thailand	Observed DDT impact on behavior of malaria mosquitoes.

[94] Slooff, R., *Observations on the effect of residual DDT house spraying on behaviour and mortality in species of the Anopheles punctulatus group. Final report on a research project in West New Guinea* (Leyden, Holland: A.W. Sythoff, 1964), 144pp.

[95] Shalaby, A.M., *Ann Entomol Soc Amer*, vol. 59, no. 5 (1966): 938–44.

[96] Smith, A. and Webley, D.J., "A verandah-trap hut for studying the house-frequenting habits of mosquitoes and for assessing insecticides. III. The effect of DDT on behaviour and mortality," *Bull Entomol Res*, vol. 59 (1969): 33–46.

[97] Ismail, I.A.H., Notananda V., and Schepens, J., "Studies on malaria and responses of *Anopheles balabacensis balabacensis* and *Anopheles minimus* to DDT residual spraying in Thailand." World Health Organization. 1975. WHO/MAL/75.847.

[98] Nutsathapana, S., et al., *Bull Ent Res*, vol. 76 (1986): 303–12.

Researchers	Study Dates	Study Location	Findings
Roberts, D.R., et al.[99]	1990	Brazil	Hut studies in a remote area of the Amazon Basin showed malaria mosquitoes stopped entering, exiting, or taking blood meals inside a hut after it was sprayed with DDT. Level of protection, compared to an unsprayed control hut, was greater than 95 percent. Authors attributed protection to true DDT repellent action, not contact-irritant or toxic action.
Fernandez-Salas, I., et al.[100]	1993	Mexico	As a result of repellent and irritant actions of DDT residues, host selection patterns switched to non-human hosts, and numbers of malaria mosquitoes entering houses dropped immediately after houses were sprayed. Study documented repellent and irritant actions.

[99] Roberts, D.R., and Alecrim, W.D., "Behavioral responses of *Anopheles darlingi* to DDT-sprayed house walls in Amazonia," *Bull of PAHO* vol. 25, no. 3 (1991): 210–17.

[100] Fernandez-Salas, I., et al., "Host selection patterns of *Anopheles pseudopunctipennis* under insecticide spraying situations in southern Mexico," *J Amer Mosquito Control Assoc* vol. 9, no. 4 (1993): 375–84.

Researchers	Study Dates	Study Location	Findings
Chareonviri-yaphap, T., et al.[101]	1997	Belize	As a result of laboratory and field tests, authors concluded that "irritant and repellent responses of malaria vectors to insecticides are important components of malaria control operations." Authors observed that insecticide resistance did not seem to alter the repellent and irritant responses of the malaria vector to the insecticide.
Casas, M.C., et al.[102]	1998	Mexico	Documented action of DDT on behavior of malaria mosquitoes.
Roberts, D.R., et al.[103]	2000	Based on data from Brazil, Suriname, Kenya and Indonesia	Modeled mosquito behavior in test huts sprayed with DDT, defining the three modes of action.

[101] Chareonviriyaphap, T., et al., "Pesticide avoidance behavior in *Anopheles albimanus*, a malaria vector in the Americas," *J Amer Mosquito Control Assoc*, vol. 13, no. 2 (1997): 171–83.

[102] Casas, M.C., et al., *J Amer Mosquito Control Assoc*, vol. 14, no. 4 (1998): 410–20.

[103] Roberts, D.R., et al., "A probability model of vector behavior: effects of DDT repellency, irritancy, and toxicity in malaria control," *J Vec Ecol*, vol. 25 (2000): 48–61.

Researchers	**Study Dates**	**Study Location**	**Findings**
Grieco, J.P., et al.[104]	2000	Belize	Study compared DDT with deltamethrin in test huts and shows strong repellent action of DDT, and strong irritant but no repellent actions of deltamethrin.
Roberts, D.R., et al.[105]	2003	Bethesda, MD, U.S.	Presented analyses of a computer simulation model on interactions of repellent, irritant, and toxic actions of DDT residues. Authors report that repellent and irritant actions must function below thirty percent efficacy before toxicity could become the primary mode of DDT action against malaria mosquitoes. Authors report that field studies show repellent and irritant actions function at levels far above thirty percent efficacy.

[104] Grieco, J.P., et al., "A comparison study of house entering and exiting behavior of *Anopheles vestitipennis* (Diptera: Culicidae) using experimental huts sprayed with DDT or deltamethrin in the southern district of Toledo, Belize, C.A.," *J Vec Ecol*, vol. 25 (2000): 62–73.

[105] Roberts, D.R., "What is the role of insecticide resistance in the re-emergence of major arthropod-borne diseases?" Institute of Medicine report, *Forum on Emerging Infections: Issues of resistance: Microbe, vectors, and the host* (Washington, DC: The National Academies Press, 2003), 94–106.

Researchers	Study Dates	Study Location	Findings
Potikasikorn, J., et al.[106]	2005	Thailand	Spatial repellency and contact irritancy demonstrated in laboratory studies using field caught anophelines.
Grieco, J.P., et al.[107]	2007	Bethesda, MD, U.S. & Thailand	Highly replicated laboratory studies and experimental hut studies demonstrate DDT acts as a first order repellent, a second order irritant, and only lastly as a killing agent.

Because DDT was toxic, they assumed toxicity was its only important chemical action. As a consequence, they designed their studies to define DDT's impact on malaria vectors, not to distinguish between different modes of chemical action. However, even in the earliest studies, the chemical's powerful repellent action on mosquito behavior was apparent.

Investigators reported on a large set of systematic studies conducted by several agencies and research groups in the Tennessee Valley during 1943 and 1944. They treated houses and barns and monitored resting mosquitoes outside before and after buildings were sprayed. In their routine house-spraying experiments, they reported that "no adult [malaria mosquitoes] . . . were ever found resting in any of the houses following treatment, although they were examined frequently during the

[106] Potikasikorn, J., et al., "Behavioral responses to DDT and pyrethroids between *Anopheles minimus* species A and C, malaria vectors in Thailand," *Am J Trop Med Hyg*, vol. 73, no. 2 (2005): 343–49.

[107] Grieco, J.P., et al., "A new classification system for the actions of IRS chemicals traditionally used for malaria control," PLoS One, vol. 2, no. 8 (2007): e716. doi:10.1371/journal.pone.0000716.

remainder of the summer and early fall."[108] In surveying residents of sprayed houses, they found that "the mosquito situation had been improved for considerable periods of time in all the houses . . . In almost all cases it was found that the mosquitoes were not staying in the houses or biting to any great extent. A typical comment was that the mosquitoes would enter the house in the evening, buzz around a while, and then leave without biting."[109]

A section entitled "Repellency of DDT-treated surfaces" was included in Metcalf and coauthors' final published work. From studies of sprayed and unsprayed barn stalls and sprayed and unsprayed nail kegs (resting sites for mosquitoes),[110] they found that the malaria mosquito would not rest on DDT-treated surfaces. In experimental house tests, they reported the efficacy of DDT in controlling malaria mosquitoes. Unfortunately, there were no counts of mosquitoes repelled without physical contact versus those irritated following contact with DDT. As a consequence, the researchers' observations on repellent versus irritant actions of DDT are not supported by hard data. The researchers found that spraying of nail kegs, barns, and houses produced the same result—a high level of irritant and/or repellent action. The authors characterized this repellent action more as contact irritancy versus non-contact repellency, but they had no count data to support their interpretations. In each case, though, the presence of DDT almost completely eliminated resting mosquitoes from inside the sprayed kegs, barns, and houses.[111]

---

[108] Metcalf et. al., 1945. 759

[109] Ibid.

[110] A nail keg is a cylindrical wooden container used for the shipment and storage of nails or other dry goods.

[111] Ibid., 761-765.

USDA entomologists, James Gahan and Arthur Lindquist, conducted another study in Arkansas in 1944.[112] They monitored areas with all buildings heavily treated with DDT (208 mg/ft^2), an area with lightly treated buildings (56 mg/ft^2), and an area with no treated buildings. The authors reported that most mosquitoes that made contact with DDT in unscreened buildings flew out open windows or doors and escaped before being knocked down or killed—that is, there were few if any dead or moribund mosquitoes in the treated buildings.

Gahan and Lindquist measured the relative numbers of adult mosquitoes in treated versus untreated areas by placing nail kegs near but outside the buildings. The kegs were open on one end so mosquitoes could enter, land, and rest. A count of resting mosquitoes over time from these protected resting sites provided a measure of what was happening with mosquito populations in and around the buildings. The malaria mosquito would congregate and rest in buildings, and it would also congregate and rest in nail kegs. So, if building treatments were actually killing the adult mosquitoes as they congregated in large numbers indoors, then there should be a rapid and steep decline in the numbers of adult mosquitoes in areas surrounding sprayed buildings. This decline would be revealed in numbers congregating in nail kegs. Gahan and Lindquist found no meaningful differences in numbers of mosquitoes resting in kegs around sprayed buildings compared to unsprayed buildings, so the nail-keg survey showed that numbers of adult mosquitoes in the surrounding areas were not reduced by DDT treatments. Yet, the numbers of mosquitoes resting in buildings were reduced by 99 percent in heavily treated buildings and by 91 percent in lightly treated buildings.

If this study had been an isolated, one-of-a-kind event, one could dismiss it, but it was not. Although the methods of field

[112] Gahan, J.B. and Lindquist, A.W., "DDT residual sprays applied in buildings to control *Anopheles quadrimaculatus*," *J Econ Entomol*, vol. 38, no. 2 (1945): 223–230.

studies vary widely, a careful reading of the literature reveals that in study after study, year after year, vector after vector, and country after country, similar results were obtained. Among the numerous studies, one in the United States, conducted in 1944 and published in 1947, is of enormous relevance.

Two U.S. government scientists with the Communicable Disease Center, Clarence Tarzwell and Frank Fisk, conducted a study of malaria mosquitoes in DDT-treated and untreated buildings.[113] They observed that sprayed walls were active (would kill mosquitoes after 20 minutes or more of physical contact) against the mosquitoes for at least eleven months. Data from their study showed that DDT greatly reduced numbers of malaria mosquitoes resting in buildings (see fig. 3-1).

Figure 3-1: Comparison of DDT Sprayed and Unsprayed Buildings

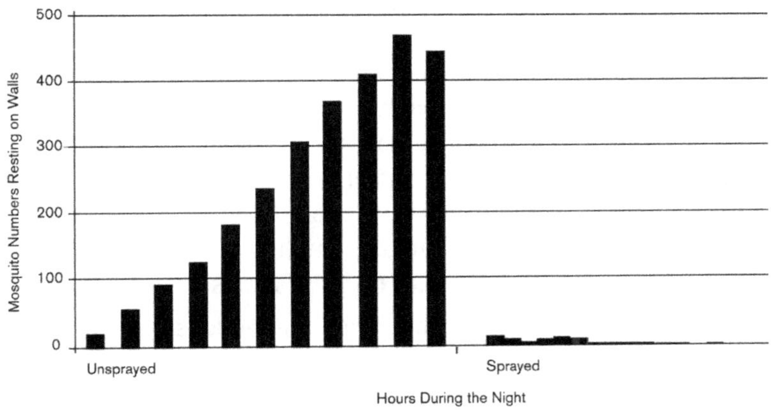

Tarzwell, C.M. and Fisk, F.W., "Observations on the nighttime resting and biting habits of anopheline mosquitoes in DDT-treated and -untreated buildings," Publ Hlth Rep, vol. 62 (1947): 84–94.

In collections over two nights, 1,997 unfed females were found resting in an untreated barn, whereas in a treated barn only

[113] Tarzwell, C.M. and Fisk, F.W., "Observations on the nighttime resting and biting habits of anopheline mosquitoes in DDT-treated and -untreated buildings," *Publ Hlth Rep*, vol. 62 (1947): 84–94.

59 unfed females could be found. Remember, these counts tally the number of resting mosquitoes, not the number of dead mosquitoes. Also remember DDT won't kill the mosquito unless it rests on sprayed surfaces. Tarzwell and Fisk also showed that

> in treated buildings the mosquitoes quickly become irritated, so that their resting period is reduced to a matter of minutes. Since no anopheline [malaria-carrying] mosquitoes and only one of the culicines [non-malaria-carrying mosquitoes] remained beyond the 15-minute intervals, it is assumed that the maximum resting period is less than 15 minutes.[114]

These researchers showed that DDT was highly effective; but it was effective not by killing mosquitoes, but by greatly reducing the number of mosquitoes resting on a sprayed wall and by reducing the resting time of the few that did land on treated surfaces. Thus, at the very beginning of DDT usage, Tarzwell and Fisk provided a clear demonstration that the complex repellent and irritant actions essentially preempted any impact from DDT toxicity. Furthermore, they properly credited these actions by stating, "The reduction of the number of resting mosquitoes in a treated building . . . might well be one reason for the protection afforded by DDT residual sprays."[115]

In laboratory research, the behavioral actions of DDT were even harder to ignore than in the field studies described previously. An internationally recognized ethologist, John Stodart Kennedy, performed laboratory studies to quantify repellency as a mode of DDT action in a 1947 paper.[116] Kennedy's experiments demonstrated the repellent and irritant actions of DDT against the malaria vector *An. maculipennis atroparvus* and the urban vector of the dengue and yellow fever

[114] Ibid., 91–92.

[115] Ibid., 90.

[116] Kennedy, "The excitant and repellent effects of mosquitoes."

viruses, *Ae. aegypti*. His work was published four years before the first report of DDT resistance in a malaria vector. He wrote,

> "During the war it was necessary to concentrate on the practical development of DDT as an insecticide without waiting for studies of its mode of action."[117]

Kennedy then presented data showing that DDT residues, at sublethal concentrations, strongly repelled mosquitoes (see fig. 3-2) and that specimens often recovered completely after showing preliminary symptoms of poisoning. With this documentation, he concluded that "DDT must be regarded as acting in two contradictory ways simultaneously. It acts as a lethal agent on the one hand, and as an excitant and thereby sometimes a repellent on the other."[118] Figure 3-2 shows the percentage of *An. maculipennis atroparvus* females escaping from DDT-treated and untreated chambers in just three minutes after first being released in treated and control (untreated) chambers. The percentage remaining accounts for those that did not immediately escape from the exposure chamber.[119]

[117] Ibid., 593.

[118] Ibid., 605.

[119] Ibid.

Figure 3-2: Percent *An. maculipennis atroparvus* escaping from chambers with sub-lethal doses of DDT vs. untreated chambers

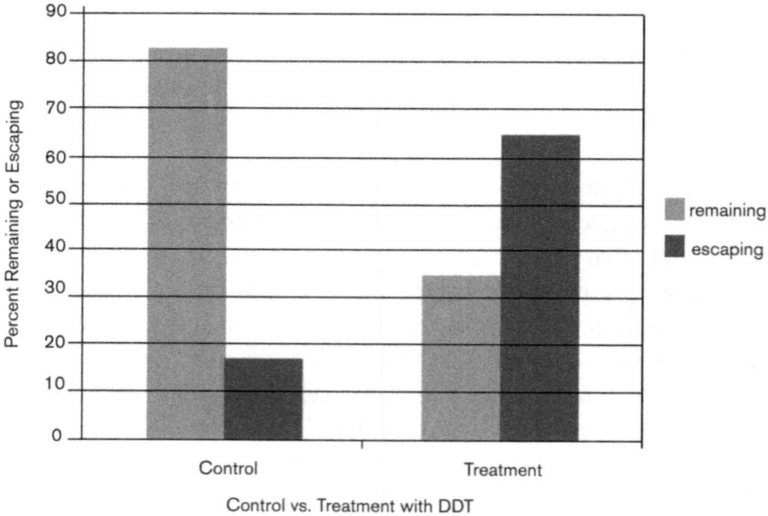

Kennedy, J., "The excitant and repellent effects of mosquitoes of sub-lethal contacts with DDT," Bull Entomol Res, vol. 37 (1947): 593–607.

Kennedy conducted his work while Macdonald was establishing the mathematical foundations of malaria control. He disagreed with Macdonald's contention that DDT functioned by killing mosquitoes. Kennedy proposed that there was both "destruction and repulsion of the mosquitoes."[120] A study by research entomologists, Richard Fay and Elizabeth Sheppard, published in 1949, confirmed Kennedy's observation that mosquitoes did recover following knockdown with DDT and that DDT provoked powerful behavioral responses in mosquitoes.[121] Equally important, Fay and Sheppard also

[120] Ibid.

[121] Fay, R. and Sheppard, E., "*Anopheles quadrimaculatus* activity patterns in the laboratory on untreated and DDT-treated surfaces," *J Nat Mal Soc*, vol. 9, no. 2 (1949): 147–58. Knockdown refers to a quick insecticide action that immobilizes the mosquito without actually killing it. With time the mosquito might recover and fly away.

showed that the mosquito must be in contact with DDT for greater than twenty minutes for more than fifty percent mortality. (Tarzwell and Fisk showed that their mosquitoes rested for less than fifteen minutes on DDT residues).

Arnoldo Gabaldon, A DDT Pioneer

Of the malaria-control specialists in the 1940s, none had greater operational knowledge or international respect than Dr. Arnoldo Gabaldon. Gabaldon's reputation among malariologists grew slowly through his work to combat malaria in Venezuela. He became director of Venezuela's national malaria-control program in 1936, where he waged an unrelenting war against the disease.

From 1936 to the mid-1940s, the primary means of attacking malaria was through large drainage projects to eliminate malaria mosquito habitats. In other words, drainage systems were constructed to dry out wetlands and other bodies of standing water. Such projects were expensive and became cost-effective only when the drainage areas were adjacent to large human populations. It was not until the advent of DDT in the mid-1940s that Gabaldon could begin to attack malaria in rural areas. His control program was the focus of a paper he presented before a meeting of the Royal Society of Tropical Medicine and Hygiene in July 1949.[122] From this paper, it is clear that he moved aggressively to incorporate the use of DDT into the national program. Gabaldon's use of DDT in Venezuela was an enormous success.[123] It greatly reduced malaria and largely eradicated one of the major malaria mosquitoes, *An. darlingi*,

---

[122] Gabaldon, A., "The Nation-wide campaign against malaria in Venezuela. Part I," *Trans Roy Soc Trop Med Hyg*, vol. 43, no. 2 (1949): 113–32.

[123] Gabaldon, A., "The Nation-wide campaign against malaria in Venezuela. Part II," *Trans Roy Soc Trop Med Hyg*, vol. 43, no. 2 (1949): 133–64.

from significant geographical areas of the country. Venezuela reported 817,115 malaria cases in 1943, but with use of DDT, only 800 cases were reported in 1958.[124]

In his presentation before the Royal Society of Tropical Medicine and Hygiene in July 1949,[125] Gabaldon was very pointed in his descriptions of DDT's repellent actions. He explained that "as soon as DDT is sprayed anophelines disappear from the houses. This may be the consequence of what seems to be its repellent action."[126] He also noted that "DDT seems, at least in some cases, to be a repellent, a chemical means of exclusion. The repellent effect of DDT is hard to explain."[127] Gabaldon may have had a hard time explaining DDT's repellent effect, but he knew it was there. He noted that while anopheline populations disappeared from sprayed houses, the mosquitoes continued to congregate in neighboring unsprayed houses. He also reported that when one spraying squad didn't spray the walls according to the regulations, but instead simply dusted the floor of the sleeping rooms with wettable DDT powder, the mosquitoes stayed away.[128]

---

[124] WHO, "Official Records No. 190, WHO, Executive Board Forty-Seventh Session," Geneva, Jan. 19–29, 1971. Part II. Appendix 14, p. 178.

[125] Gabaldon, "Venezuela. Part II," 141.

[126] Ibid., 147.

[127] Ibid., 155.

[128] Some may consider that DDT would direct mosquitoes away from sprayed houses to unsprayed houses where malaria transmission could continue. However it is vital to note that Gabaldon does not state that mosquitoes move from sprayed houses to unsprayed houses, just that normal movement of mosquitoes to the unsprayed houses continues unabated. Furthermore not all houses in Gabaldon's program were sprayed (some were excluded because residents were away or because they refused entry to the spraymen) and yet malaria cases fell impressively. The evidence Gabaldon presents along with evidence gathered from the long history of DDT and malaria control around the world demonstrates that the idea that DDT would divert mosquitoes to

Gabaldon stressed that "this repellent action, however, should not be confused with the effect apparently produced by DDT in some species, which may inhibit the mosquitoes from remaining on sprayed surfaces, forcing them to leave the house."[129] The importance of his observation relates to two actions of DDT on mosquito behavior: a contact-irritant action and a non-contact, spatial-repellent action. These terms describe a chemical action to which the mosquito has a specific response. A spatial repellent is a chemical that induces directed movement away from a chemical source without physical contact. A contact irritant is one that causes locomotor stimulation or agitation following physical contact with the chemical. In other words, the chemical makes the mosquito break contact and move away from the chemical residue. Toxicity generally occurs after physical contact with the toxic compound and can take the form of death, rapid knockdown, or knockdown followed by death.

Through his observations, Gabaldon made an original, but largely unappreciated, contribution to the science of malaria control. By distinguishing between two behavioral actions, he became the first malaria-control worker to draw a distinction between the chemical actions of non-contact repellency and contact irritancy. He correctly attributed different roles to these actions, describing the former as "exclusion" and the latter as "interception."[130] To describe Gabaldon's terminology, exclusion means the directed movement of a malaria mosquito away from a chemical source without physical contact. In other words, the mosquito detected DDT from a distance and moved away without ever making physical contact with the substance. This is our definition of spatial repellency. In practical terms, exclusion would prevent the mosquito from entering the house. Interception, however, describes a mosquito being so irritated by

---

unsprayed houses and that malaria transmission would continue has no basis in fact.

[129] Gabaldon, "Venezuela. Part II," 155.

[130] Ibid.

the chemical following physical contact that it quickly leaves the area. This is our definition of contact irritancy. In practical terms, then, interception would allow the mosquito to enter the house, but would also cause the mosquito to prematurely exit the house.

Gabaldon's insight into the chemical actions of DDT was out of step with the beliefs of many program managers, researchers, and public health bureaucrats in the malaria-control community. As luck would have it, one of the discussants who attended the 1949 presentation before the Royal Society of Tropical Medicine and Hygiene was Macdonald. Macdonald was impressed that Gabaldon's programs were eradicating *An. darlingi* from areas in Venezuela, but thought that although the DDT did appear to have a repellent action, this characteristic compromised malaria eradication, saying that "in Africa, the discovery that DDT repels *A. gambiae* more than it kills was a setback to hopes of eradication by this means."[131]

In those early days of DDT use, Macdonald's opinion carried great weight and his views were adopted and repeated by many in the malaria-control community. He is credited with establishing the widely held belief that DDT killed malaria mosquitoes and, as a result, reduced overall mosquito population longevity (average number of days adult female mosquito lives) below the number of days for completing development of malaria parasites in the mosquito. Thus, on average, the females did not live long enough to become infectious and transmit disease. Belief that this was how malaria was controlled was deeply entrenched. Due to the strength of this belief, any mechanism of chemical action that did not result in mosquito death was often considered undesirable. Without ever evaluating the benefit versus harm of DDT's repellent actions, many malariologists simply adopted the view that the repellent action would undermine malaria control. However, others did not accept this view and, as we will show, they went on to

[131] Ibid., 161.

demonstrate that the repellent and irritant actions did indeed serve to exert important levels of control over malaria.

Macdonald was a man of enormous stature and intelligence, but in this case, he just didn't get it. Perhaps Gabaldon's respect for Macdonald's international stature caused him not to respond in the meeting with the points he made in his paper. Gabaldon never stated that killing mosquitoes was the means of eradicating *An. darlingi* from some areas of Venezuela. Instead, he stated that when houses were sprayed, the mosquitoes disappeared from the houses. In his presentation, Gabaldon clearly understood that he was discussing very complex chemical, behavioral, toxicological, ecological, and epidemiological interactions. It seems likely that he also understood that experimental data for understanding these complex interactions were entirely inadequate. Perhaps for those reasons, he admonished his audience that the full measure of DDT effectiveness should be based on DDT's impact on malaria, not impact on mosquitoes. In spite of this admonition, as revealed by Macdonald's comment, the focus remained on interactions of DDT and mosquitoes, not DDT and malaria.

These two great intellects of malaria control were juxtaposed at the 1949 meeting. One was forged through a career of mathematical modeling and buoyed by immense international stature. The other was forged through the grueling work and heavy responsibilities of operational disease control and characterized by keen insight into the complex issues of malaria-control operations. In the international arena, stature won the day and the views of the former prevailed, setting the stage for the colossal public-health disaster to come.

Gabaldon also called *An. darlingi* the most domesticated of all of Venezuela's malaria mosquitoes. This meant that it was the most endophagic and most anthropophagic: it aggressively entered houses to feed on humans. However, when a house was sprayed with DDT, the chemical's repellent action stopped it from entering houses. The mosquito would then decline in abundance, *not because it was being killed by DDT, but because the DDT was keeping it away from the crucial substance it needed to reproduce: blood.*

An. darlingi is a forest mosquito strongly associated with rivers and streams. It becomes abundant only in the presence of humans, because humans are its optimal host. Furthermore, as long as humans are present, this species can thrive in otherwise unfavorable environments. Given these factors, it is intuitively reasonable that when denied access to humans in otherwise unfavorable environments, *An. darlingi* could disappear completely from such locations. This appears to have happened in some areas of Venezuela and other peripheral areas of the Amazon Basin. Gabaldon expounded on such ecological factors in a later publication.[132]

If our interpretations and Gabaldon's observations are correct, then a failure to use DDT would allow *An. darlingi* populations to return and proliferate in all their old haunts. As it happens, this is precisely what occurred.[133] Where DDT had lowered *An. darlingi* populations below detectable levels, those populations resurged with discontinued use of DDT.[134] With more mosquitoes came increased malaria.[135]

This is only one example of malaria vector elimination from geographic areas by spraying houses with DDT.[136] Indoor

---

[132] Gabaldon, A., "The effect of DDT on the population of anopheline vectors in Venezuela," *Riv Parassitol*, vol. XIII, no. 1 (1952): 24–41.

[133] Roberts, D.R., et al., "Role of residual spraying for malaria control in Belize," *J Vector Ecol*, vol. 27, no. 1 (2002): 63–69.

[134] Guarda, J.A., et al., "Malaria reemergence in the Peruvian Amazon Region," *Emerg Inf Dis*, vol. 5, no. 2 (1990): 209–15; http://www.cdc.gov/ncidod/eid/vol5no2/aramburu.htm (accessed Apr. 12, 2009).

[135] Roberts, D.R. et. al., "DDT, global strategies, and a malaria control crisis in South America," *Emerg Inf Dis*, vol. 3, no. 3 (1997): 295–302.

[136] Our use of the word "elimination" does not mean true elimination because, in most cases, the vectors were only reduced to such low densities as to be undetected in routine surveillance efforts.

spraying also eliminated *An. funestus* in parts of Tanganyika and South Africa.[137] When DDT use was stopped in South Africa, populations of *An. funestus* reappeared after many years of absence.[138] DDT eliminated *An. minimus* from areas of Thailand and *An. darlingi* from the coastal regions of Guyana and Suriname. Use of DDT in Peru reduced *An. darlingi* populations below detectable levels and maintained control over malaria for many years. When spraying was stopped there in the late 1980s, populations of *An. darlingi* became more common and malaria reemerged as a major public-health problem in Peru.

The 1950s: The Beginning of Global Malaria Eradication

The 1940s literature shows that field reports about DDT accumulated quickly during the early years of its use in malaria control, and the repellent/irritant actions of DDT were a common theme in those reports. Yet despite those reports, there was a strong allegiance to Macdonald's model and the belief that DDT functioned by killing mosquitoes. Other malariologists, supported firmly by their field experiences, wrote against this popular belief, and researchers continued to study DDT's mode of action.[139]

Working with the Indonesian Ministry of Health in 1951, Pieter Hendrik van Thiel attempted to create a structure of

---

[137] Gillies, M.T. and Smith, A., "The effect of a residual house-spraying campaign in East Africa on species balance in the *Anopheles funestus* group. The replacement of *A. funestus Giles* by *A. rivulorum Leeson*," *Bull Entomol Res*, vol. 51: 243–52.

[138] Hargreaves, K. et. al., "*Anopheles funestus* resistance to pyrethroid insecticides in South Africa," *Med Vet Entomol*, vol. 14, no. 2 (June 2002): 181–89.

[139] Muirhead-Thomson, R.C., *Mosquito behaviour in relation to malaria transmission and control in the tropics* (London, Edward Arnold & Co., 1951). 219 pp.

different categories of behavioral responses of malaria vectors to DDT. Van Thiel stated that he had a problem visualizing the repellent effect.[140] Yet, in work with two species, *An. sundaicus* and *An. subpictus*, he observed that it was hard to find dead mosquitoes in sprayed houses. Other researchers shared van Thiel's experience. Van Thiel observed that *An. sundaicus* disappeared immediately from sprayed villages, while *An. subpictus* was more persistent. He found an average of forty malaria mosquitoes per day for several days in an experimental house. After the house was sprayed, he found scarcely any mosquitoes, dead or alive. He concluded that some form of repellency could hardly be denied, and presented other data and reports from other studies to support his conclusion, which agreed with Gabaldon that DDT exhibited strong behavior-modifying actions.[141]

Just two years later, in 1953, Macdonald and Davidson (a research medical entomologist with the Ross Institute) published what became the founding principles for using DDT in malaria control.[142] They based their analyses on field studies from 1947 to 1952. They did not cite Kennedy's paper,[143] and they dismissed the relevance and importance of any behavioral responses to DDT that did not result in mosquito deaths. Because they focused on killing mosquitoes, Macdonald and Davidson struggled valiantly to reconcile DDT's slow killing

[140] The low vapor pressure of DDT caused confusion about its repellent effect. Workers assumed that because DDT has a low vapor pressure, it would leave too few molecules in air to allow mosquitoes to detect the chemical and respond with repellent behavior. For that reason, workers more readily accepted the idea of behavioral avoidance following physical contact than behavioral avoidance without contact.

[141] Van Thiel, P., "The repellent effect of insecticides, especially of contact insecticides," *Doc Neerl Indones Morb Trop*, vol. 3 (1951): 117–25.

[142] Macdonald and Davidson, "Dose and cycle of insecticide application."

[143] Kennedy, J., "The excitant and repellent effects of mosquitoes."

power with its spectacular ability to reduce malaria. The authors knew of DDT's complex influence on vector behavior, noting several times that DDT had a "marked irritant effect."[144] They concluded that the answer to DDT's "irritant" action was to increase dosage so that mosquitoes would absorb a lethal dose before chemical repulsion.

Their conclusion was an unfortunate lapse of logic. If Macdonald and Davidson had paid attention to Kennedy's paper, they would have understood that DDT's behavioral actions were not linked to toxicity. Indeed, sublethal concentrations of DDT stimulated the repellent actions reported by Kennedy. Kennedy reported that for contact-irritant actions, "excitation appears quickly, often in a matter of seconds after the insects are brought into contact with DDT."[145] So, the logical conclusion should have been that increasing the dose will not alter the chemical actions or the insect's responses to those actions since the actions and responses occur at sublethal concentrations in the first place. What was unknown, and remained unknown until recently, was how weak the concentration could be while still provoking behavioral actions in malaria mosquitoes. Based on recent research of medical entomologists at the Uniformed Service University of the Health Sciences at Bethesda, Maryland, we now know that a repellent response can be provoked by DDT concentrations 200-fold less than the

[144] Macdonald and Davidson, "Dose and cycle of insecticide application," 808. Macdonald and Davidson further explained on page 792 that "DDT was the most irritant of the three insecticides [DDT, BHC, and Dieldrin], and a very large proportion of the mosquitoes escaped its action." On page 798 they state that "The marked irritant properties of DDT were evident in all these treatments, the greater proportion, and sometimes as much as 90 %, of the mosquitoes being caught in the window traps.", and on page 808 add that "The marked irritant effect of DDT on mosquitoes makes adequate dosage . . . imperative."

[145] Kennedy, "The excitant and repellent effects of mosquitoes."

operational dose recommended for spraying on house walls to kill mosquitoes.[146]

The global malaria-eradication program began in 1955, just two years after Macdonald and Davidson published their model. Accepted without question, their model became a core component of the global program. Their assessment that DDT killed mosquitoes and reduced mosquito longevity below levels for completing extrinsic incubation of the malaria parasites was wrong, but it completely dominated the doctrine of malaria-control operations.

The 1960s: The Beginning of the End for DDT

The appearance of DDT resistance and the formal startup of the global program helped change research emphasis from testing the efficacy of DDT in control programs to resistance testing and monitoring in spray programs. Remarkably, there was never a coordinated effort to define DDT's mode of action, yet WHO did support researchers who dutifully reported the repellent actions of DDT residues. Rudolf Slooff was one such scientist. From July 1959 to January 1963, he conducted a particularly insightful study on DDT actions in New Guinea.[147] Slooff observed that fifty percent fewer malaria vectors (specimens of *An. punctulatus*—the primary vector species) would enter DDT-sprayed houses compared to unsprayed

[146] Achee, N., Sardelis, M.R., Dusfour, I., Chauhan, K.R., and Grieco, J.P., "Characterization of spatial repellent, contact irritant, and toxicant chemical actions of standard vector control compounds." J Am Mosquito Control Assoc, 2009, 25(2):156-167.

[147] Slooff, R., *Observations on the effect of residual DDT house spraying on behaviour and mortality in species of the Anopheles punctulatus group. Final report on a research project in West New Guinea* (Leyden, Holland: A.W. Sythoff, 1964) 144 pp. Slooff's work in New Guinea was supervised by Professor van Thiel and supported by WHO.

houses. His writing showed that he understood that he was observing a very important, perhaps dominant, aspect of DDT's function when sprayed on house walls.[148] Furthermore, those mosquitoes that entered the sprayed houses exited much more quickly, and a smaller proportion engorged on blood. Regarding his observation that far fewer mosquitoes entered sprayed versus unsprayed houses, Slooff wrote, "The mosquitoes must . . . have been affected at some distance from the DDT-sprayed huts."[149]

In this report, Slooff described a non-contact repellent action that prevented fifty percent of malaria mosquitoes from entering sprayed houses. However, at that time, mainstream belief held that toxic action was the sole means of malaria control, and because it was politically unwise to talk about DDT as a repellent, Slooff omitted the repellent action from his discussion and summary.[150] At one point in the results section, Slooff even stated, "Although there would grammatically be no objection

---

[148] Slooff wrote, "Various explanations have been proposed for this phenomenon [repellent action of DDT residues]. E.g., Downs & Bordas (1951), Zulueta et al. (1961), Kuhlow (1962) and Kuhlow & Garms (1962) suggested that the DDT inside their test huts exerted some repellent effect, and that this prevented part of the mosquitoes from entering. The same assumption was made by Wilkinson (1951) . . .

Although we must do without an insight in its causes, we are obviously dealing with a phenomenon of importance." Slooff, *Observations on residual DDT house spraying*, 58–59.

[149] Slooff, *Observations on residual DDT house spraying*, 61.

[150] After reading Slooff's dissertation, one of the authors (Donald Roberts) attended a conference in Woods Hole, MA, in November 1993, where he spoke with Slooff and told him that his data showed clearly that DDT was functioning as a non-contact repellent against vectors in New Guinea. Roberts pointed out that Slooff described this repellent action in the results section of his dissertation but hardly mentioned it in the discussion or in the summary. Roberts asked Slooff why he had not elaborated on this in the discussion and summary, and Slooff responded that during those years it was politically unacceptable to talk about DDT as a repellent.

against the use of the word repellent for this effect, we will refrain from it."[151]

There were other prominent malariologists who tried to change opinions about the importance of DDT's complex actions against malaria. Julian de Zulueta was one such malaria expert who, by the end of 1960, had researched DDT's actions in several countries and against different malaria mosquitoes. While working within WHO's Division of Malaria Eradication in Geneva in 1963, Zulueta published a summary of his findings.[152] He announced that DDT did not function to control malaria by killing mosquitoes:

> DDT, it was assumed until recently, interrupts malaria transmission by its lethal effects, by the reduction in numbers or the total eradication of the vector. From recent work on irritability, it appears that this may not be the case. After enforced contact with DDT and after having shown marked signs of irritation, only very few mosquitoes are found dead 24 hours after exposure. DDT seems to kill few mosquitoes and to drive most of them away. It thus seems to act more by exclusion than by reduction.[153]

Zulueta also suggested that other prominent malariologists, Mario Coluzzi and Robert Muirhead-Thomson,[154] had come to

---

[151] Slooff, *Observations on residual DDT house spraying*, 61.

[152] Zulueta, J., "The irritability of mosquitoes to DDT and its importance in malaria eradication," *Riv Malariol*, vol. XLI (1962): 169–78.

[153] Ibid., 169–70.

[154] Coluzzi, A. and Coluzzi, M., "Sull'irritabilita degli *Anopheles* al DDT in rapporto all'eradicazione della malaria," *Riv Malariol*, vol. 40 (1961): 35–40; Muirhead-Thomson, R.C., "The significance of irritability, behavioristic avoidance and allied phenomena in malaria eradication," *Bull Wrld Hlth Org*, vol. 22 (1960): 721–34.

the same conclusion that irritation was an important facet of DDT's action.

By the end of 1960, many years of extensive DDT use in malaria control had passed, and reports had accumulated on DDT resistance. The view remained strong that DDT resistance meant that DDT would no longer be effective in affected areas. However, Zulueta differed from mainstream opinion; he did not believe that resistance meant that DDT was ineffective.[155] Zulueta, referring to his past work, stated,

> [A]fter investigating in Greece the resistance of *sacharovi* and the behaviour pattern of this species, [Zulueta] came to the conclusion that irritability accounted for the slow development of DDT-resistance, the physiologically susceptible mosquitoes being driven away from the treated surfaces before having picked up a lethal dose, thus rendering inoperative what would have been a selective action of the DDT. The result of this is that, after ten years of DDT-resistance, DDT still remains the most effective insecticide against malaria in Greece. Dieldrin and BHC cannot be used due to the widespread development of resistance of this type in the local *sacharovi*, but DDT by its irritant effect, by driving away the susceptible fraction of the *sacharovi* population, reduces the contact with man and, as a consequence, the chances of transmission.[156]

[155] However, Zulueta, along with most malariologists of that era, did seem to believe that physiological resistance to insecticides and behavioral avoidance of the same insecticide were inextricably linked. This perspective is hard to understand because resistance and behavioral avoidance are expressed through very different mechanisms. It seems quite reasonable that they are under entirely different and independent genetic functions.

[156] Zulueta, "The irritability of mosquitoes to DDT," 175–76.

Note Zulueta's assessment that dieldrin resistance rendered the chemical ineffective; but that DDT resistance did not. We will return to this topic later.

On DDT's irritant actions, Zulueta summarized his views in a strong statement:

> Against the general judgment, that the irritant effect of DDT is a disadvantage in the use of this insecticide, the author believes, with others, that irritability might have a beneficial effect by driving away the mosquitoes entering treated dwellings, thus reducing the contact with man and achieving, as a consequence, the interruption of transmission.[157]

In 1960, the global eradication program was in full force and DDT was being used in many malaria-endemic countries around the world. There was great momentum behind spray programs and global monitoring of insecticide resistance.[158] But, like studies by many other scientists, Zulueta's strong statements had little impact on the popular opinion that DDT functioned by killing mosquitoes.

At the time of Zulueta's 1960 paper, he had initiated a pilot study to evaluate DDT for malaria eradication in Uganda. The study would give Zulueta a firsthand experience with another aspect of DDT's action on mosquito behavior. The first spray cycle was conducted in Uganda in mid-1959, and Zulueta demonstrated that once houses were sprayed, the numbers of mosquitoes inside houses dropped dramatically,[159] while numbers in unsprayed shelters remained high, at least through

---

[157] Ibid., 176.

[158] Coincidentally, 1960 saw the beginning of global efforts to find toxic chemicals as substitutes for DDT.

[159] Zulueta, J., et. al., "The results of the first year of a malaria eradication pilot project in northern Kigezi (Uganda)," *E Afr Med J*, vol. 38, no. 1 (1961): 1–26.

the first six months. Afterward, numbers in outdoor shelters also declined. This gradual decline probably reflects a slow drop in the population density of malaria vectors as they were denied access to their optimal host: humans. Zulueta described DDT's impact on malaria and malaria mosquitoes as "remarkable."[160] He reported that in a sprayed hyperendemic area,[161] malaria dropped from a prevalence of 22.7 percent of persons examined to just 0.5 percent in only ten months. In a mesoendemic area,[162] prevalence declined from 12.5 percent to 0 percent.[163] Zulueta did not find a lot of dead mosquitoes on floors, however. He clearly described a DDT mode of action that is different from the contact-irritant action. Specifically, he described a chemical action that prevents mosquitoes from entering sprayed houses—in other words, a spatial-repellent action. Based on his data and figures, he stated that

> the impression may be gained that . . . DDT reduced the number of *A. gambiae* by about two-thirds, when in fact, if the repellent effect of the insecticide is taken into account, it will be seen that it reduced the number of *A. gambiae* leaving the hut by eight-ninths approximately. This great reduction is in accordance with the complete interruption of transmission and the great drop in parasite rates observed throughout the protected area.[164]

---

[160] Ibid., 15.

[161] Gillies, H.M. and Warrell, D.A., *Bruce-Chwatt's Essential Malariology* (London: Edward Arnold, 1993), 131. A hyperendemic area is one that features intense, seasonal transmission where immunity is insufficient to prevent the effect of disease on all age groups.

[162] Ibid. A mesoendemic area is one where the disease is found among small, rural communities with varying intensity depending on local circumstances.

[163] Zulueta et al., "Malaria eradication pilot project," 21.

[164] Ibid.

Intrigued by the results of the pilot study, Zulueta conducted experimental hut studies, which he published in *Nature* in 1963, to quantify DDT actions that reduced numbers of malaria mosquitoes entering houses.[165] In these experimental hut studies, he clearly discriminated between irritant and repellent actions. He reported that DDT reduced hut entry (compared to unsprayed huts) by 80 to 88 percent. Referring to his work and that of others, Zulueta stated that the chemical action that prevented hut entry was not an irritant action. He also concluded that the chemical action preventing hut entry must have great significance for malaria-control operations. As a result of his research, Zulueta fully embraced the idea of three separate DDT actions: toxic, contact irritant, and non-contact repellent.[166]

The strong position taken by Zulueta and co-author Cullen brought forth a rebuttal published in *Nature* by Alec Smith,[167] who was involved in WHO's global search for DDT substitutes.[168] Smith concluded that the data by Zulueta and Cullen did not constitute proof that the malaria vectors were greatly deterred from entering treated huts. Yet, Smith's data could neither prove nor disprove Zulueta's work. To Smith's great credit, he then investigated DDT's repellent and irritant actions with studies in the Magugu area of Tanzania and in the Taveta area of Kenya from 1964 to 1966. In these areas, Alec Smith and his co-author D.J. Webley worked with two of the most important malaria vectors in Africa, *An. gambiae* and *An.*

---

[165] Zulueta, J. and Cullen, J.R., "Deterrent effect of insecticides on malaria vectors," *Nature*, vol. 200 (1963): 860–61.

[166] Zulueta describes non-contact repellency as a deterrent action.

[167] Smith, A., "Deterrent effect of insecticides on malaria vectors (rebuttal)," *Nature*, vol. 200 (1963): 861–62.

[168] Dichlorvos is an organophosphate compound that is widely used as an insecticide. It is most commonly used in enclosed spaces where its fumigant action can kill insects in stored products. Dichlorvos is not recommended by WHO for use in indoor residual spraying for malaria control.

funestus.[169] They used experimental huts and published their work in 1969.[170] They found that sixty to seventy percent of *An. gambiae* "were deterred from entering verandah-type huts treated indoors with DDT, for at least four months after treatment. In the Taveta area, *An. funestus* was less deterred from entering than *A. gambiae*."[171] They further reported that "the DDT deposit also had a marked irritant effect on *A. gambiae* and *A. funestus*."[172] Smith and Webley admitted that Zulueta and Cullen had been correct; DDT functioned as both a repellent and an irritant.[173] This was a critically important paper, but as with all those that had come before, it was ignored. It was not published in a highly visible journal, so it was not widely read. Those who read the 1963 article in *Nature* would not learn that Smith subsequently agreed with Zulueta and Cullen.

Still, for the malaria mosquitoes in Africa, these findings were far from the first. Responses of *An. funestus* to DDT were actually described in research conducted in Mauritius and

---

169 Both species are extremely important vectors of malaria. Both species are highly endo- and anthropophagic. Both are widely distributed in Africa. *An. gambiae* is considered the most efficient and dangerous vector of human malaria in the world.

170 Smith, A. and Webley, D.J., "A verandah-trap hut for studying the house-frequenting habits of mosquitoes and for assessing insecticides. III. The effect of DDT on behaviour and mortality," *Bull Entomol Res*, vol. 59 (1969): 33–46.

171 Ibid., 43; It is important to note that *An. funestus* was repelled by DDT, just seemingly less so in this test than was *An. gambiae*. Previous citations show that indoor spraying can suppress *An. funestus* populations below detectable levels and stop malaria transmission.

172 Smith and Webley, "A verandah-trap hut," 45.

173 Smith and Webley also observed the amounts of DDT moving out the house as a vapor and the dose on individual mosquitoes that resulted in death. These statistics are discussed later in this chapter (see "How Mosquitoes Detect DDT in the Air?").

published in 1948.[174] The paper was cited in a book published in 1951.[175] The book's author, Muirhead-Thomson, described the studies with *An. funestus* and stated unambiguously that "despite the high *funestus* population before treatment, there was an almost complete absence of dead mosquitoes inside the treated houses—a finding which was attributed to the fact that D.D.T. must exert some repellent effect, preventing *funestus* from entering houses."[176] These studies represent just a few of the many replicated studies confirming that the popular belief about DDT was wrong—it did not function by killing mosquitoes.

Given the high frequency and consistency of reports and observations on DDT's repellent and irritant actions, the persistence of the popular belief that DDT functioned in malaria control solely by killing mosquitoes is a striking testimony to the power of a popular belief over science. The systematic studies of Zulueta and Cullen, of Smith and Webley, and all the others were just ignored. Popular belief that DDT functions by killing mosquitoes is wrong, but it persists to this day as an official guiding policy of the World Health Organization.

Many official documents contain evidence of this persistent view. For example, in the 1995 synthesis of conditions for indoor spraying, a WHO study group surmised that "indoor residual spraying can reduce the vector life span, [and] vector populations."[177] In its 2001 recommendations for DDT use inside houses for malaria control, WHO staff characterized indoor spraying as follows: "Vectors may avoid the effect of an insecticide sprayed indoors. The term 'behavioural resistance' has been used to describe the way that vectors survive in sprayed

---

[174] Muirhead-Thomson, R.C. *Mosquito behaviour in relation to malaria transmission and control in the tropics* (London, Edward Arnold & Co., 1951), 102.

[175] Ibid.

[176] Ibid.

[177] WHO, "Vector control for malaria and other mosquito-borne diseases." WHO Technical Report Series 857, Geneva, 1995, p. 20.

areas by avoiding sufficient contact with the sprayed surfaces."[178] The study group went on to say that "indoor spraying is ineffective when the vector does not come into the house for biting or resting. This may . . . be because . . . of 'excito-repellency,' which is when the insecticide causes the vector to avoid it."[179] It defined excito-repellency as being "important in vector control because its effect is to divert the vector from resting on sprayed surfaces and, therefore, from acquiring a lethal dose of the insecticide."[180]

Furthermore, for the purpose of spraying walls, the organization stated in 2002 that "it is important to recognize the essential characteristics of spraying when choosing it as a control intervention. The expected result of indoor residual spraying is mainly to reduce the survival of vector(s) entering houses."[181] The same document identified indoor spraying as the primary means of increasing adult vector mortality, stating that "increasing adult vector mortality reduces their expectation of life and therefore the probability that the parasite will complete its development."

These WHO staff members identified DDT as the most effective insecticide, but there was no meaningful recognition that DDT did not function by killing mosquitoes. Additionally, they did not list indoor spraying as a means of breaking human-vector contact inside houses because they emphasized killing mosquitoes to the practical exclusion of any other beneficial actions of DDT residues.[182]

---

[178] Najera, J.A. and Zaim, M., "Malaria vector control: Insecticides for indoor residual spraying." WHO, 2001, WHO/CDS/WHOPES/2001.3, p. 37.

[179] Ibid., 37.

[180] Ibid., 58.

[181] Ibid., 2.

[182] Ibid., 12–17.

The Trumpeting of Resistance Problems

As soon as insecticides became commercially available, farmers almost immediately snapped them up and began applying them extensively. This widespread use of insecticides in agriculture rapidly selected for resistance in mosquitoes and other insects because when any insecticide (DDT, malathion, pyrethroids, etc.) is sprayed in the environment, insects often cannot avoid it. Therefore, mosquitoes and other species tend toward resistance to agricultural insecticides, even though these insecticides are intended for agricultural pests.

Because DDT was broadly used in agriculture, there was considerable selective pressure for DDT resistance in mosquitoes. Consequently, the reports of vector resistance to DDT (as well as to other insecticides, such as pyrethroids), continued to grow. In spite of all countering evidence on how DDT functioned to control malaria, the focus of most malaria-control research through the 1960s was on the killing actions of DDT and its alternatives. Thus, reports of vector resistance propelled the malaria-control community to a state of despair.

The concerns about resistance and the misunderstandings about the way in which DDT works clearly hampered malaria control. For instance, the Sri Lankan malaria-eradication program used DDT to reduce malaria cases to an all-time low of just eighteen cases in 1963, five years after the DDT-spraying campaign started. After the five-year attack phase, the program was cut because it was assumed that DDT resistance would render the insecticide useless.[183] This, for many unfortunate Sri Lankans who had suffered from malaria for countless generations, was a fatal mistake. Studies showed that *An. culicifacies*, a major vector, was not only wholly susceptible to DDT, but that DDT was working well.[184] In fact, the question of

[183] Spielman and D'Antonio, *Mosquito*, 173.

[184] Harrison, G., *Mosquitoes, malaria and man. A history of hostilities since 1880* (London: John Murray, 1978), 255.

the vector's susceptibility to DDT should have been secondary to whether DDT spraying still served to control malaria. The repellent and irritant actions might have still functioned to control disease even in the face of resistance to DDT's toxic actions. Continued spraying with DDT might well have delayed or even prevented the upsurge in malaria cases. However, public-health officials continued to base decisions for replacing DDT in house-spray programs on insecticide-resistance test data alone.[185] As stated in a 2001 WHO manual, "An essential characteristic in the choice of an insecticide is the susceptibility of the malaria vectors."[186] Yet, despite WHO's search for a DDT substitute (screening mostly agricultural products for efficacy against mosquitoes), no one had identified any cheap, long-lasting DDT substitutes for use in malaria control.

All the while, pressure from environmentalists to abandon the use of public-health insecticides was gaining political momentum. Smith and Webley's important paper on DDT's modes of action went unnoticed, but even if it had been noticed, WHO's global malaria-eradication program had begun to wane and was formally ended by the Twenty-Second World Health Assembly in 1969.[187] WHO staff in the Division of Malaria and Parasitic Diseases in Geneva (no longer called the Division of Malaria Eradication) said the malaria-eradication program failed because it "had not succeeded in outstripping the development of insecticide-resistance in the vectors."[188] Anopheline resistance, that "had been mainly induced by the contamination

---

[185] Brown, A.W.A., et al., "Review Article. Malaria eradication and control from a global standpoint."

[186] Najera and Zaim, "Malaria vector control: Insecticides for indoor residual spraying," 36.

[187] WHO, "Re-examination of the global strategy of malaria eradication," WHO official records, 1969, no. 176, annex 13, 106–26.

[188] Brown et al., "Review Article. Malaria eradication and control from a global standpoint," 11.

of the larval breeding areas with insecticides applied to agricultural crops, especially cotton," destroyed the program.[189]

After many years of malaria-control efforts, adherents to the idea that DDT controlled malaria by killing mosquitoes had come full circle. They had initiated the global eradication program because of fear that mosquito resistance to DDT's toxic actions would make it ineffective in the battle against malaria. They wanted to use DDT quickly to eradicate malaria before resistance became a bigger problem. Then, years later, they claimed the program was stopped because they believed resistance had become an insurmountable problem. Missing in this whole sequence of events was the recognition that DDT didn't even function through toxic actions on mosquitoes. Also missing from such assessments was any recognition that the global eradication program was actually stopped for complex ideological reasons that had nothing at all to do with the minor technical issues of DDT resistance. In chapter 6 we will provide in depth coverage of the ideologies that ultimately ended malaria eradication.

How DDT Actually Functions to Control Malaria

Malaria vector mosquitoes enter houses, bite humans indoors, and exit in a genetically programmed sequence of behaviors. Depending on the time of day, or, more accurately, on the photoperiod (hours of daylight or hours of darkness), DDT's various actions of repellency, irritancy, and toxicity can intervene at specific points in the mosquito's sequence of behaviors. For instance, repellent action can prevent the mosquito from entering the house altogether. Contact-irritant action can prevent biting by stimulating a premature exiting behavior in those that do enter.[190]

---

[189] Ibid.

[190] However, a negative aspect of the contact-irritant action would be to stimulate premature exiting after the mosquito bites, but before it absorbs a toxic dose of DDT. Roberts, D.R., et al., "A probability

Lastly, toxic action can intervene after the mosquito has entered and rested on surfaces with DDT residues.

In simple terms, the separate actions of DDT residues can provide an interactive and cumulative benefit. For the purposes of illustrating these interactions, imagine that repellent, irritant, and toxic actions each are seventy percent effective. If one hundred mosquitoes enter a house not sprayed with DDT, then only thirty would enter if it were treated with DDT, as the other seventy would be repelled by the DDT and not enter. Of the thirty mosquitoes that enter the house, twenty-one would be stimulated by contact-irritant actions and exit. This leaves nine mosquitoes that remain indoors and bite. Of these, six or seven would absorb a toxic dose of DDT and die, and the other three or four would escape and survive. By summing the numbers of mosquitoes from the three interventions, we find that although all of these actions function at an equal level of activity (seventy percent), spatial-repellent and contact-irritant actions together reduce the overall risk of being bitten and acquiring disease by nintey-one percent, while toxicity reduces risk by only six to seven percent. Overall risk would be reduced by nintey-five to nintey-six percent.

---

model of vector behavior: Effects of DDT repellency, irritancy, and toxicity in malaria control," *J Vec Ecol*, vol. 25 (2000): 48–61; Interactions described above involve the multiplication law of probabilities, if **A** and **B** are two events, the probability of their joint occurrence, i.e., P(**A** and **B**), is P(**A** and **B**) = P(**A**) P(**B**|**A**). This is the general case of the multiplication law.

Figure 3-3: Depiction of mosquito behavior in unsprayed house for 100 mosquitoes

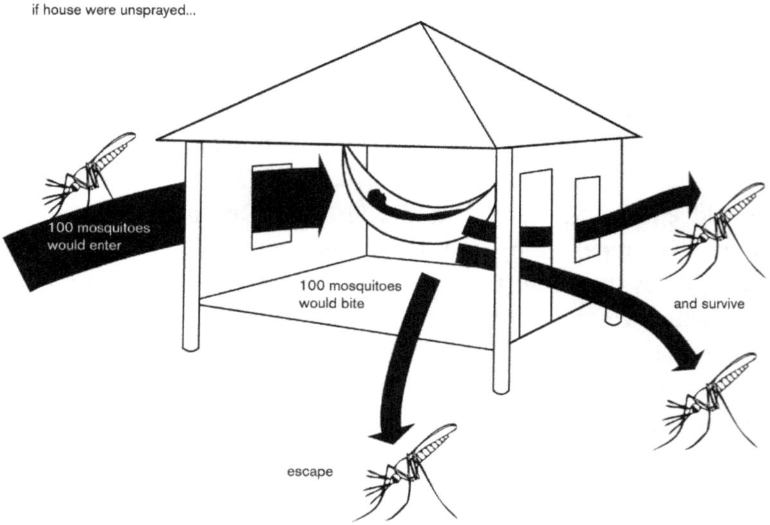

Figure 3-4: Depiction of mosquito behavior in DDT-sprayed house for 100 mosquitoes

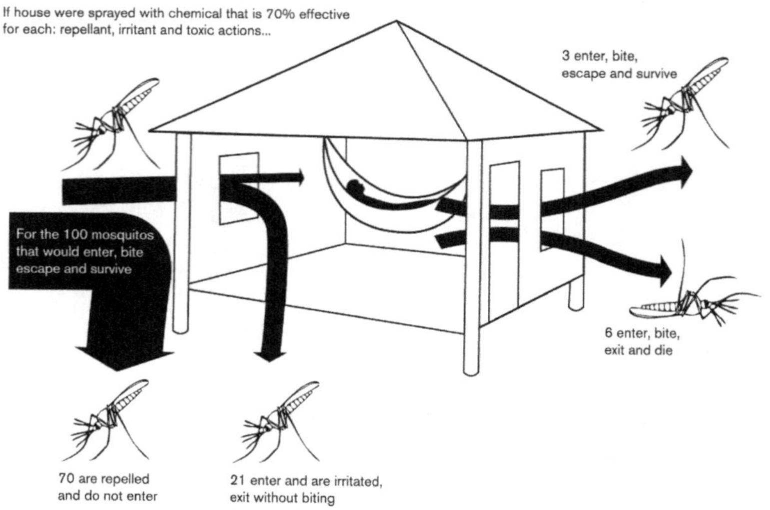

This example is based on repelling and irritating only seventy percent of the mosquitoes. Computer simulations of interactions between repellent, irritant, and toxic actions show that an insecticide would need to repel and irritate less than thirty percent each before toxicity would become the primary mode of chemical action against malaria vectors.[191] Yet, in the real world, levels of DDT's repellent and irritant actions are far above seventy percent. Of the insecticides presently recommended for malaria control, only DDT has these three chemical actions. For this reason, no other insecticide is as effective against most vector mosquitoes in most epidemiological settings for malaria control.

Figure 3-5: 3 typical Mayan houses.

[191] Roberts, D.R. and Hshieh, P., "What is the role of insecticide resistance in the re-emergence of major arthropod-borne diseases," Knobler, S.L. et. al., eds., *The Resistance Phenomenon in Microbes and Infectious Disease Vectors: Implications for Human Health and Strategies for Containment – Workshop Summary* (Washington, DC: Institute of Medicine, 2003), 94–104.

A field study from southern Belize illustrates the astonishing differences between DDT and one alternative insecticide.[192] Many residents of southern Belize are native Americans and live in typical Mayan houses.

The study compared DDT with the pyrethroid insecticide, deltamethrin, in three typical Mayan houses with thatch roofs and wood slat walls. The malaria vector in question was *An. vestitipennis*, and mosquitoes were captured as they entered and exited the huts, both before and after spraying. One hut was sprayed with DDT, another was sprayed with deltamethrin, and the third hut was not sprayed at all. The results showed clearly that DDT stopped most mosquitoes from entering the hut (Figure 3-6). Mosquitoes readily entered the deltamethrin hut and there was some biting during the evening hours. However the mosquitoes became irritated by the insecticide and exited prematurely (Figure 3-7). In contrast to the unsprayed hut, mosquitoes in the deltamethrin hut were all gone by midnight. These reactions to insecticides are in marked contrast to mosquitoes' normal pattern of entering and biting. Thus, both chemicals influenced mosquito behavior, but only DDT stopped the malaria mosquito from entering the huts. As shown in the graphs, the mosquitoes remain and bite inside unsprayed huts throughout the entire night.

192 Grieco, J.P., et al., "A comparison study of house entering and exiting behavior of *Anopheles vestitipennis* (Diptera: Culicidae) using experimental huts sprayed with DDT or deltamethrin in the southern district of Toledo, Belize, C.A." *J Vec Ecol*, vol. 25 (2000): 62–73.

Figure 3-6: DDT vs Deltamethrin. Entrance times of *An. vestitipennis* compared to control and pre-spray

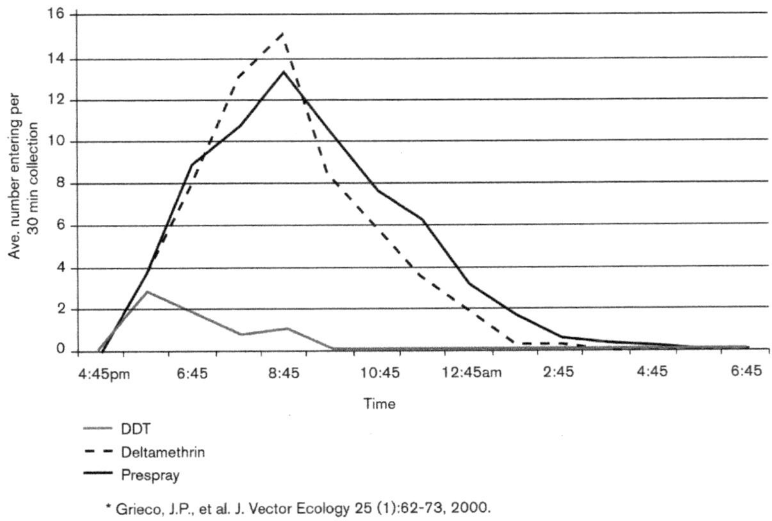

Figure 3-7: DDT vs Deltamethrin. Exit times of *An. vestitipennis* compared to control and pre-spray

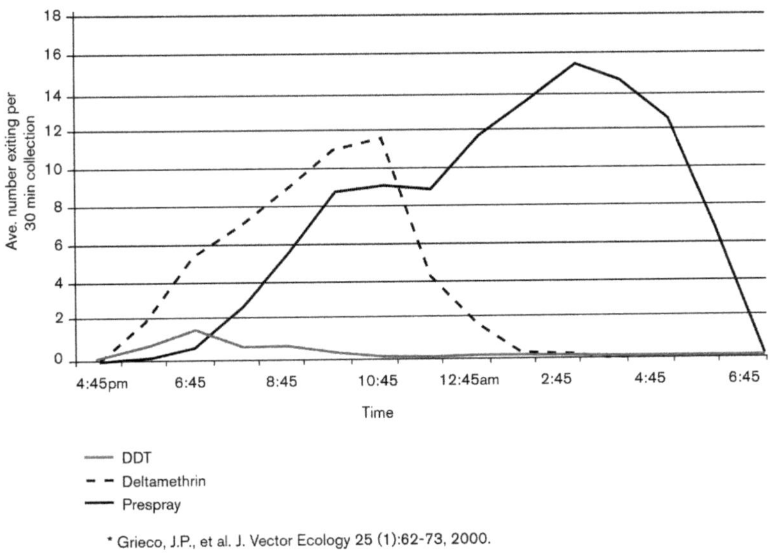

How Mosquitoes Detect DDT in the Air

DDT has a very low vapor pressure, so only low concentrations of DDT molecules will be found in the air. Initially, this fact led to assumptions that DDT could not influence mosquito behavior unless it made physical contact with sprayed surfaces. However, the number of molecules in a unit of air is not as important as the mosquito's sensitivity to a specific chemical. Sensitivity encompasses a capability of the mosquito to detect and respond to a gradient of increasing numbers of molecules in the air. Detecting a gradient of increasing numbers of molecules tells the mosquito from which direction the chemical originates and allows it to move away from the chemical source. Based on vapor-pressure calculations, the saturation of indoor air with DDT would roughly equate to 0.3 nanograms of DDT per liter of air, or 0.0000000003 grams of DDT per liter of air ($3x10^{-10}$ grams). While it might seem improbable that a mosquito could detect such minute quantities of a chemical, in fact insects can detect far more dilute quantities of chemicals than this. In one instance, a male silk moth was found to detect the female sex attractant at 200 molecules per ml of air, i.e., one odor molecule among 10^{17} molecules of air.[193] To put these quantities in perspective, proportionately, this would be a single teaspoon of chemical in 13 million Olympic-sized swimming pools, each containing 3,750,000 liters of water. However, the specific question that we need to answer is whether it is possible for malaria mosquitoes to detect very small quantities of DDT.

As shown in the review of historical literature presented earlier, scientists answered this question decades ago. In 1969, Smith and Webley, who had conducted extensive research in Tanzania on the repellent properties of DDT, provided

---

[193] Schoonhoven, L.M., "Insects in a Chemical World," 1-21, 3, in *CRC Handbook of Natural Pesticides. Volume VI. Insect Attractants and Repellents.* Eds. Morgan, E.D. and Mandava, N.B, CRC Press Inc. Boca Raton, FL, 1990

particularly insightful data.[194] They calculated concentrations of DDT in indoor air, quantified DDT residues on sprayed and unsprayed surfaces, and monitored the movement of DDT out of huts. Their research showed that minute quantities of DDT in air stopped most malaria mosquitoes from entering. Smith and Webley also provided insight into the concentrations needed for toxic versus behavioral actions. They estimated that 7 to 20 ng of DDT on or in the mosquito caused death, while they found that DDT would repel mosquitoes at concentrations ranging from 0.4 to 0.08 ng per liter of air in the sprayed huts. While Smith and Webley did not define the minimum concentration that would repel mosquitoes, they clearly demonstrated the huge differences in the quantity of DDT needed for the different actions. The midpoints for the two sets of data, 13.5 ng of DDT in or on the mosquito for toxicity and 0.24 ng per liter of air for repellency, show the larger quantities needed to induce toxic responses compared with the minute quantities needed to induce spatial-repellent responses.

Given the consistency, reproducibility, and solid documentation of the behavioral actions of DDT, how has WHO's strategy for malaria control or its strategy for discovering new chemicals for malaria control taken this evidence into account? How has WHO incorporated this information into its search for alternative insecticides or in evaluating the effectiveness of DDT in different geographical locations, against different vectors, and in different epidemiological settings? Sadly, the answer is that it has not. The true mode of DDT action does not fit at all in WHO's thinking or strategies.[195] As WHO's 2002 manual shows, WHO still adheres to the belief that the main action of DDT is toxic.

---

[194] Smith and Webley, "A verandah-trap hut."

[195] We describe how WHO factors the spatial-repellent and irritant actions into its guidelines for indoor spraying near the end of the section, "The 1960s: Beginning of the end for DDT".

The Proportional Dose–Response Relationships of DDT Use and Malaria Cases

The history of attempts to understand how DDT works has sometimes been complex. Ultimate evidence that it works is found in documentation that there is a true proportional dose–response relationship between spraying houses and controlling malaria. Although this relationship is complex, the fundamentals are relatively straightforward.

A threshold level of effective spraying for a given human population is the number of houses that must be sprayed to prevent more cases of malaria. This threshold can be empirically defined by fitting historical house-spray and malaria data to mathematical models.[196] The models can determine the minimum proportion of houses that must be sprayed in order to prevent malaria from increasing. When spraying is below a threshold, then malaria cases will increase in proportion to the reductions in numbers of houses sprayed. When spraying is above that threshold, then the numbers of cases will decline in proportion to the increase in numbers of houses sprayed. The proportionality of these relationships is also expressed over time. In other words, a sudden large reduction in spraying will result in a rapid increase in cases, and gradual reductions in spraying will result in gradual increases in cases.

These relationships were tested for Brazil, Ecuador, and Belize.[197] The thresholds, which were based on real data, were

[196] Specifically, these were logistic regression models that regressed a binary variable for increasing or decreasing malaria against the numbers (or proportions) of houses sprayed (used the standard house spray rate values for this independent variable).

[197] Logistic regression models were tested against thirty-four years of malaria-control data (Pan American Health Organization, "Status of Malaria Programs in the Americas. XLII Report." PAHO, 1994, Washington, DC,) for Brazil, Ecuador, and Belize. For these models, and other analyses in this report, published methods were used to standardize annual parasite indexes (APIs), which express the yearly

52 houses sprayed per 1,000 population in Brazil, 58.4 for Ecuador, and 134.6 for Belize.[198] In other words, if fewer than 52 houses per 1,000 population are sprayed in Brazil, then malaria rates will increase. Spray programs in Brazil, Ecuador, and Belize dropped permanently below threshold spraying in 1975, 1980, and 1990, respectively.[199] Correspondingly, malaria increased almost every year after a country program dropped below its specific threshold. Rapid drops in spraying and rapid increases in disease occurred in Belize and Peru, while Brazil slowly reduced spraying and saw gradual increases in malaria.[200]

---

number of slide-diagnosed malaria cases per 1,000 population (Roberts D.R., et al., *J Vector Ecol* 2002, 27(1):63–69). Logistic regression models developed and tested with SAS software. A derived binary variable was the response variable and the house-spray rate was the independent variable. A binary variable was developed for each country by assigning a value of 0 if the standardized API increased during the following year and 1 if it declined. The API for each country was standardized against the average annual blood-examination rate for years from 1965 to 1979. Significance was established at the 0.05 probability level.

[198] A statistically significant ($p<0.05$) fit of data to the models defined a MEHSR of 52 houses sprayed per 1,000 population for Brazil, 58.4 for Ecuador, and 134.6 for Belize (Roberts D.R., et al., 2002. *J Vector Ecol* 27(1):63–69). All three models illustrate the proportional dose–response relationship between malaria incidence and levels of indoor DDT spraying. Brazil, Ecuador and Belize dropped permanently below MEHSRs in 1975, 1980, and 1990, respectively (Pan American Health Organization, "Status of malaria programs in the Americas. XLII Report." PAHO, Washington, DC, 1994).

[199] Pan American Health Organization, "Status of malaria programs in the Americas. XL Report." (Washington DC: PAHO, 1991).

[200] Data on the number of houses that needed to be sprayed to eradicate malaria evolved during the late 1940s and 1950s. During the global eradication program, WHO recommended that 80 percent of houses be sprayed in order to completely stop malaria transmission. Locations where 80 percent coverage did not completely stop malaria transmission, even if it did reduce transmission, became known as refractory areas. See WHO, 21st World Health Assembly, "Report on

Threshold levels of spraying are strongly influenced by the proportions of total populations at risk of malaria. For example, Belize, where almost everyone is at risk of malaria, had the highest threshold.[201] However, fewer houses need to be sprayed if historical and environmental data are used to more carefully select houses for spraying. The reason is that, historically, all houses in a malaria control area are sprayed. However, it is well known that all households are not at equal risk of malaria transmission even though reasons for such differences are not well understood. If spraying is targeted by risk assessments (usually based on history of malaria infections in individual households), then fewer houses can be sprayed and still obtain a satisfactory level of malaria control. This type of control has been described and even used, to a limited extent, in Belize with good results.[202]

The actual meaning of the threshold value is also important. For Belize, the threshold is 134.6 houses per 1,000 population. However, according to WHO guidelines, each house is sprayed twice per year. So the actual number of individual houses sprayed is really 67.3 houses. If we expect 5 members per household, then a population of 1,000 will be housed in 200 houses. If 67.3 must be sprayed to prevent malaria increases, then 33.6 percent of all houses in Belize need to be sprayed. Thirty-three percent seems like a lot, but remember, the whole population of Belize is at risk of malaria. If we apply the same statistics to Brazil, where only a small proportion of the population is at risk of malaria, we find that only 13 percent of houses need to be sprayed and almost all will be located in the

---

development of malaria eradication program", Geneva, Mar. 13, 1968, for a discussion of the eradication program and those areas where malaria resurged where transmission was not stopped.

201 Roberts, D.R., et al., "Role of residual spraying for malaria control in Belize, C.A.: A case study." *J Vector Ecol*, 1999, 27(1):63–69.

202 Ibid.

Amazon Basin, where malaria remains endemic. These analyses show the dose–response relationships of IRS.

The relationships between DDT use and malaria control and the existence of true dose–response relationships between proportions of houses sprayed and levels of disease within populations document why elimination of DDT from national malaria-control programs has been a catastrophic mistake—horrific consequences have resulted for poor people who are at an ever-increasing risk of catching malaria. This is illustrated in growth of malaria cases and declining use of IRS for the countries of Brazil, Colombia, Peru, Ecuador and Venezuela.

Figure 3-8: Cumulative numbers of excess malaria cases over average numbers per year (1965-1979) for Brazil Colombia, Peru, Ecuador and Venezuela.

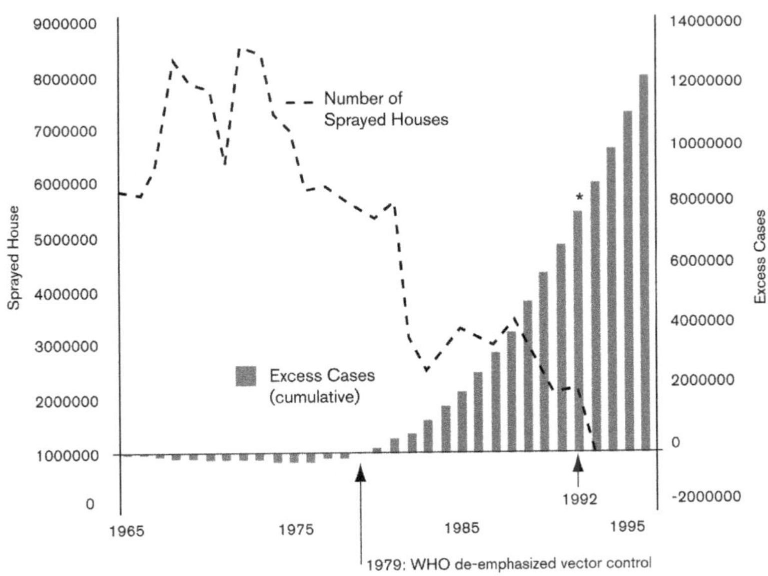

*First year excess cases grew by more than 1 million cases/annum

Data adjusted to a standerdized ABER.

However, knowing these relationships does not give us a clear understanding of why spraying was systematically reduced in the first place.

Summary

Despite high-quality and widely published scientific evidence and abundant experience from the field showing that DDT primarily acts as mosquito repellent, public-health officials in WHO as well as many malaria scientists have clung to the notion that DDT acts primarily as a toxic agent. So what? If the insecticide works, who cares how it actually functions? Yet, understanding how DDT works is, in fact, crucially important, because many DDT detractors see a limited usefulness for the insecticide because of the prospect of toxic resistance. They argue that if mosquitoes develop resistance to DDT and it no longer kills them, the chemical has no use for reducing malaria. They would be right if DDT were primarily a toxic agent, but in fact it is primarily a repellant that acts secondarily as an irritant and lastly as a toxic agent. This means it is still useful in malaria control even in the presence of resistance to toxicity. Thus, a lack of understanding of how DDT works has resulted in it not being used and lives being lost.

Chapter 4

DDT: Totemic Villain of the Environmental Movement

The 1960s was a period of significant change in the United States, Western Europe, and almost everywhere else. While the winds of change blew through Africa, starting the process of African independence from colonial rulers, students in the West rebelled against the post-war leadership. At the same time, Rachel Carson, a biologist, freelance science writer, and editor-in-chief for the U.S. Fish and Wildlife Service, published her famous book, *Silent Spring*,[203] a book that is widely seen as launching the modern environmental movement. Because of this, Carson is recognized as one of the most influential people of the twentieth century. One of those profoundly influenced by her is former vice president Al Gore, who wrote in an introduction to the 1994 edition of *Silent Spring*, "Indeed, Rachel Carson was one of the reasons why I became so conscious of the environment and so involved with environmental issues."[204] He went on to state that "Without this book, the environmental movement might have been long delayed or never have developed at all."[205]

---

[203] Linda Lear, "Rachel Carson's Biography," http://www.rachelcarson.org/Biography.aspx (accessed July 16, 2009).

[204] Carson, R., *Silent Spring* (New York: Houghton Mifflin Company, 1994), xviii.

[205] Carson, *Silent Spring*, xv.

This chapter explains the complex and often misunderstood relationship between DDT and bird populations. We have provided this level of detail precisely because the idea that DDT led to devastating harm to bird populations is so widely and unquestioningly accepted. We give this thorough analysis and presentation of data to refute claims that DDT caused catastrophic harm to birds. Additionally, we provide a more accurate and fair history because some may be unwilling to give up their long held beliefs. Regardless, some readers may find this history dense and may not wish to wade through the detail that we have provided. Therefore a brief overview of this chapter is given below.

Overview

DDT is thought to have harmed wildlife populations, but we focus on three in particular: robins, the bald eagle, and the peregrine falcon. In each of these cases, DDT was accused of harming their populations, bringing them to the brink of extinction. In each case, the banning of DDT in 1972 is thought to have saved these birds and led to their eventual revival. These ideas are false.

Rachel Carson wrote passionately, as ever, about the harm that DDT inflicted on robins when the chemical was sprayed on Michigan State University campus to halt the spread of Dutch elm disease which was harming trees on the campus. On a local level, DDT did indeed harm some robins through acute poisoning. However, Carson took this kernel of truth and claimed that DDT would spread far and wide in ever widening circles and would harm the bird's ability to reproduce. DDT in soils would accumulate in earthworms and Carson claimed that when robins fed on the worms, they would be ingesting a lethal dose of the chemical. In reality DDT did not spread far and wide and where it was applied, it degraded, posing less of a threat to the robins than when it was freshly sprayed. A careful study of robin populations in orchards in Canada found that robins were unaffected by DDT and high levels of DDT in earthworms had no effect on the robin's ability to reproduce. Furthermore, an examination of robin populations shows that their numbers did not decrease during the years that DDT was used; indeed populations increased. Robins prefer open landscapes with hedges and shrubs, such as those found in suburban gardens of America. As America's suburbs expanded, providing increased habitat for the birds, so did robin populations.

Unlike robins, bald eagles populations were indeed imperiled in the United States, but they were imperiled long before DDT was ever used. Changes in land use harmed bald eagle populations, and like other wildlife, their numbers declined because they were ruthlessly hunted, trapped and poisoned either for sport or to protect livestock. In an attempt to save the bald eagles, the U.S. Congress passed the Eagle Protection Act in 1940. Some environmental campaigners claim that this act helped to save bald eagles, only to have them further harmed when DDT was introduced after WWII. In reality the Eagle Protection Act was poorly funded and poorly enforced and was of little practical help to bald eagle populations. One significant problem was that juvenile bald eagles are indistinguishable from golden eagles, which were not protected by the Act. Therefore, hunting of juveniles continued unabated. It was not until 1962 that the Eagle Protection Act was amended to include golden eagles, which gave solid protection to these juvenile bald eagles. Because the Act was poorly enforced, hunting, trapping and poisoning of adult bald eagles also continued until the passage of the 1973 Endangered Species Act. Bald eagle numbers were, however, increasing even before this time, thanks to the protection of juveniles, and populations were recovering well. Nesting pairs doubled between 1963 and 1974. Bald eagles are now no longer endangered, thanks to improved legal protection and direct action under the ESA to reintroduce breeding pairs. The history of bald eagles in the U.S. shows us that the decline and recovery of this species is closely linked to human activities that first harmed populations and then allowed them to recover. Neither the decline nor the subsequent recovery of this bird are linked to DDT.

Another raptor supposedly harmed by DDT is the peregrine falcon. Like bald eagles, these birds were shot and frequently taken as young chicks by falconers. Additionally, falcons hunt pigeons and song birds, which gave it a very bad reputation among pigeon fanciers and anyone who valued song birds. Alarming though it may be to read this now, the 1935 edition of *Hornaday's American Natural History* encourages readers to shoot and kill male and female falcons and to take the eggs and young. A fascinating study of falcon populations of the Hudson River between 1930 and 1960 explains that hunters, egg collectors, and falconers were exploiting the falcon populations heavily. There was a respite during WWII when young men were away fighting in Europe and the Pacific and when petrol was rationed, which limited the ability of falconers to travel to eyries and exploit the birds. The decline in the fortunes of falcons resumed, however, once servicemen returned from the war, many as better marksmen, and fuel rationing ended.

Our fundamental argument in this chapter is that environmental campaigners have, perhaps through ignorance, negligence, or more likely due to their own firmly held beliefs, continuously ignored or downplayed the real and documented threats to wildlife from non-chemical factors and have trumpeted the chemical factors. While these chemical factors may theoretically have played a part, the weight of evidence on this topic points to non-chemical sources of wildlife harm.

Foundations of Environmentalism: Fear and Uncertainty

In fact, concern about the environment was not new in the United States in the 1960s. At the beginning of the nineteenth century James Fenimore Cooper lamented the passing of the wilderness in his *Leatherstocking Tales*, expressing in his writing a Romantic view of the land that became popular during the first part of that century. Once the wilderness was less of a threat, people began to appreciate its beauty. It was not until the late nineteenth century, though, that an environmental movement actually began to form, brought about by what its founders saw as the ravages of the Industrial Revolution.[206]

The conservation movement gained considerable momentum in the early twentieth century, with the presidency of Theodore Roosevelt and the strong personalities of people like John Muir and Gifford Pinchot.[207] At the same time, a public-health movement arose, which was concerned particularly about health hazards in the workplace. Doctors such as Alice Hamilton drew attention to health effects on factory workers of white phosphorous, lead, and other industrial substances and sounded alarms about the fast-developing petrochemical field, arguing that not enough was known about the health effects of those chemicals for them to be considered safe, certainly for the factory workers continuously exposed to them, but possibly for their children as well.[208]

By the 1960s, the world was well into the nuclear age, and *Silent Spring* became a pivotal and highly influential publication. The book was written at a time when insecticides

---

[206] Shabecoff, A Fierce Green Fire, 22.

[207] Gottlieb, R., *Forcing the Spring* (Washington, DC: Island Press, 2005), 20–25. As is the case now, the environmental movement (if such it can be termed) of the early 20th century was not a cohesive group, but rather a variety of factions occasionally united for a similar sort of goal. The movement split generally into two major groups—the preservationists, headed by Muir, and the conservationists, headed by Pinchot—but grouped under these two useful umbrellas were a variety of factions motivated by differing ideologies and goals. As a preservationist, Muir essentially believed that the wilderness should be preserved for its intrinsic value. Pinchot, a conservationist, believed that man could utilize the resources available in nature, but that the proper application of science and technique was the way to access those resources, rather than indiscriminate development that caused extensive and unnecessary harm to nature.

[208] Gottlieb, *Forcing the Spring*, 47–51. Hamilton championed the idea that some toxins were "reproductive poisons" that could have detrimental effects on the children of the factory workers.

such as DDT and many other synthetic chemicals were, for the first time, being used in large quantities. As a consequence, more and more people were interested and rightly concerned about the possible effects of these chemicals on wildlife and humans alike. Well-documented events such as the mercury poisoning of Minamata Bay in Japan served to increase concern for the environment and led to a growth in environmentalism.

Yet, there is an important difference in being concerned about the environment and blindly following a cause that is not based on science or fact. *Silent Spring* preached mightily against man-made chemicals and, in particular, against DDT. However, while many factors can cause changes in the natural environment, environmentalists' claims of wildlife harm focused almost exclusively on insecticides and DDT. The drive to blame DDT turned into a witch hunt that jettisoned objectivity and the requirements for scientific evidence. Even a quick reading of *Silent Spring* reveals no serious consideration of other natural or man-made factors that influence the fortunes of wildlife populations. An examination of species reportedly endangered by DDT reveals a list of species that were largely endangered by human activities long before advent of DDT. The list also reveals not one species that was thriving before DDT, then suddenly declined with the broad use of insecticide. Carson tried to make the American robin such a species. She claimed DDT was endangering it, but her claim was quickly proven false. As it turns out, if there was a species that should have been endangered by DDT (as we will explain shortly), it should have been the robin. Yet, the robin was not endangered at all.

Readers of *Silent Spring*, in the 1960s and even now, are impressed by its poetic language and imagery, but it did not escape the notice of scientists that while the book was heavy on prose, it was light on science. It seems certain that scientists and students of chemistry and the natural world could not have guessed how *Silent Spring* would pave the way for science to be sidelined in the development of laws, policies, and global strategies for disease control. To this day, *Silent Spring* is seen as a standard bearer for the environmental movement. This is a shame: so many of the claims made in the book do not stand up to scientific scrutiny. On the contrary, many legitimate concerns, such as indiscriminate and massive habitat destruction, were not served well by *Silent Spring*.

Carson wrote four books; her last, *Silent Spring*, was published in 1962. The title raises the specter of a scene of enormous devastation of birds, wildlife, and public health and captured Carson's belief that our use of insecticides was responsible for this devastation. She did not write about personal experiences, and at no point did she claim to have personally witnessed the killing of birds from use of DDT. However, she claimed that the result of using chemicals like DDT would be springs without the songs of birds and the tranquil beauty of birds in flight. The central villain in this book was DDT.

Carson's attack on DDT was based on highly selective instances of insecticide usage harming local wildlife. Then Carson extended these select examples to claim harm that went far beyond what science could support. For example, DDT's opponents cling to an erroneous but popular and often-repeated story that bird populations were severely harmed by DDT through insidious effects. This story had its beginnings in the pages of *Silent Spring*. Embellished versions of the story go on to claim that bird populations were saved by the DDT ban and that bird populations underwent miraculous recoveries once DDT was no longer in use. Next, we examine the truth of these claims.

Robins and DDT: The Whole Story

Carson's writing made the plight of the robin famous and placed this bird as a central issue in the anti-DDT campaign. Carson emphasized the devastation of the robin on the Michigan State University (MSU) campus in East Lansing, Michigan. In the 1950s and early 1960s (up to 1962), DDT was sprayed on campus to protect American Elm trees from Dutch elm disease.[209] The trees were sprayed in the spring, and sometimes again in July. A natural consequence was that some robins died from acute DDT toxicity, a price the university authorities thought was worth paying in order to save a large number of trees that adorned the campus. However, Carson was not content reporting that birds were dying from acute poisoning of DDT—she took the dangers to robin populations further.

She described events on the MSU campus by giving the reader a nugget of truth followed by wildly unfounded and fearful speculation.

> Not all robins receive a lethal dose [the nugget of truth], but another consequence may lead to the extinction of their kind [the statement inducing fear] as surely as fatal poisoning. The shadow of sterility lies over all the bird studies and indeed lengthens to include all living things within its potential range [the wildly unfounded speculation]. There are now only two or three dozen robins to be found each spring on the entire 185-acre campus of Michigan State University, compared with a conservatively estimated 370 adults in this area before spraying.[210]

There is no uncertainty here: Carson was implying that DDT was causing the extinction of the American robin. For Carson, what was done on the MSU campus embodied an evil, insidious force. She concluded that spraying DDT led to "something more sinister—the actual destruction of the birds' capacity to reproduce."[211] Yet Carson was wrong, and her claims are false.

---

[209] Dutch Elm Disease is a fungal disease spread from tree to tree by the elm bark beetle. The DDT spraying targeted the beetle, thereby halting the spread of the fungus.

[210] Carson, *Silent Spring*, 108.

[211] Ibid.

There is no dispute that DDT was used on the MSU campus. Spraying was a judgment about the need to preserve elm trees versus the losses of a few robins on campus. Ironically, Carson argued with conviction that the demise of elm trees was man's own fault because we planted them to exclusion of other trees.[212] She seemed to argue that in contrast to the elm trees, whose abundance was spurred by man, robins should not be harmed because they are products of the natural world. In fact, the burgeoning populations of robins were a product of human activities. This was as true in the 1950s as it is today, as illustrated by population trends of wintering robins in Washington, D.C.

As stated in a recent article,[213] surveys of wintering birds in Washington, D.C., began in 1942. Robins wintering there were

---

[212] Carson states, "A generation or more ago, the towns of large areas of the United States lined their streets with the noble elm tree. Now the beauty they hopefully created is threatened with complete destruction as disease sweeps through the elms, carried by a beetle that would have only limited chance to build up large populations and to spread from tree to tree if the elms were only occasional trees in a richly diversified planting." Carson, *Silent Spring*, 10.

[213] In 2004, Paul DeAnna wrote, "Data for Washington's Christmas Count goes back to 1943. Not until 1950, when 10 were found, is the American Robin first recorded. Twice during the 1950s, in '55 and again in '57, the number jumps to over 200 birds, but in some years falls back to zero; the average is around 20 or so. In the '60s the average drops to just over ten birds but, as the '70s begin, there is a slow, steady increase, and by the 1980s we are into triple digit numbers. In 1990s, however, things really get going: the yearly average for the "go-go" decade is a whopping 476! Based on my first pass at the data, then, it did seem that the bird's winter numbers were up significantly."

The author goes on to explain the increase in robins, as follows: "As the 20th century progressed, the economy grew and so did the U.S. population; mid-century saw a post-war boom (remember the decade the robins were first reported?). With this expansion came an increase in the need for housing and, as a result, wooded areas were increasingly cleared for suburban developments. Where tall trees had first reported in the 1950s, and the numbers increased from that point forward. There is a difference between winter and summer populations, but the trends are similar: both populations have increased. Increases accompanied the growing pace of clearing tall trees to make room for lawns and ornamental shrubs. Robins, which prefer cleared areas, thus benefit greatly from the urbanization of America. Thus, the large populations of both robins and Dutch elms are products of human activities. However, unlike robins, the number of elms has declined. From a population of over 5,000 trees on the MSU campus prior to arrival of Dutch elm disease in the 1930s and 1940s, only four or five hundred remain today.

The period when DDT was used to protect trees was a time of experimenting and learning. The methods of fighting Dutch elm disease had not fully evolved by the early 1950s, and it remains a legitimate question whether any trees would have survived that early onslaught of disease on the MSU campus without spraying DDT. Basically, the university was in a desperate fight to salvage what it could. Without a doubt, some robins died from DDT toxicity (we emphasize that freshly-sprayed DDT will, when ingested in sufficient quantities, kill birds) when the trees were being sprayed in the spring, and we know that robin reproduction was impaired in sprayed areas. Impaired reproduction resulted from parent birds dying before the young fledged.

---

been, lawns and ornamental plantings grew up. Now, what bird was poised to take advantage of this American approach to prosperity? None other than the American Robin, a species that has a distinct preference for cleared areas, and which could now freely hunt these new lawns for worms and search the shrubbery for berries. In short, human changes to the landscape coincidentally resulted in an increased food supply for these birds, even though other species more at home in woodlands probably suffered as a consequence. DeAnna, P. *The American robin: American success story?* The Audubon Society of the District of Columbia, 2004, http://www.dcaudubon.org/node/704. (accessed Sept. 24, 2009).

Historical data on robin populations on MSU campus show that Carson's description of an "ever-widening wave of death that spreads out, like ripples when a pebble is dropped into a still pond"[214] following DDT was false. Proof of error in Carson's assessments and analogy is documented through a 1979 nest census and review of historical data.[215]

Contrary to Carson's fearful predictions, numbers of adult robins present before nesting in the spring actually reached record numbers on campus during years of DDT spraying. She was wrong in her claim that population densities were reduced to just two or three dozen nesting pairs compared to over 370 adult birds before the spraying program began. The 1979 study showed that there were never more than about fifteen nesting pairs on the campus. In other words, DDT caused no reduction in nesting pairs at all. Although spraying reduced nesting success by killing some adult birds while they were nesting in the spring, DDT had no impact on the number of eggs laid (clutch size) or on the hatching of eggs. Although some birds died, through recruitment (migration from other areas), pre-nesting populations actually remained constant. Carson's fear mongering of DDT as causing "something more sinister—the actual destruction of the birds' capacity to reproduce," was pure fiction. Nesting success returned to normal after spraying was stopped in 1962.

Carson's claims of DDT harm to robins went far beyond the situation on the MSU campus, however. She claimed that DDT endangered birds across the whole landscape of the United States, stating that "Like the robin, another American bird seems to be on the verge of extinction."[216] The other bird was the eagle, which we will discuss later, but there was no ambiguity in her statement that the American robin was on the verge of

[214] Carson, *Silent Spring*, 127.

[215] Beaver, D. L. "Recovery of an American robin population after earlier DDT use," *J Field Ornithol*, vol. 51, no. 3 (1980): 220–28.

[216] Carson, *Silent Spring*, 118.

extinction. As shown in data for robin populations in Washington, D.C., this claim was fantasy presented as fact. Other sources of data also attest to the same trend of increasing numbers of robins. In reality, robins were increasing in numbers throughout the United States even when she claimed they were moving toward extinction. Moreover, robin populations have continued to increase ever since.[217]

In addition to Carson's claims that DDT spraying directly threatened bird populations, one of her great successes was in helping to establish the belief that finding a chemical residue in the environment is equivalent to finding harm. This belief is evident in the enormous number of published papers that essentially report nothing more than residue levels in one environmental sample or another. Such reports almost never include descriptions of actual harm. One can only assume that this stems from the belief that a mere finding of chemical residue is harm enough. Of course it isn't: a finding of residue does not equate to harm—a finding of residue is just a measure of chemical, nothing more. One might conclude that the authors do not report harm because they think no harm occurs. If so, why expend vast quantities of public funds to monitor and report on trace quantities of chemical residue?

Carson and her acolytes make much of the issue of DDT persistence in soil. They imagine that DDT settles into soil and continues killing and harming wildlife from that point forward. Much of this belief stems from Carson's writing. In her comments about insecticide persistence in soil, she states that as "the virtually indestructible residues continue to build up in soil, it is almost certain that we are heading for trouble."[218] This statement is another example of misinformation because

[217] U.S. Geological Survey, "Audubon Christmas bird count. American robin *Turdus migratorius*," http://www.mbr-pwrc.usgs.gov/bbs/cbcgr/h7610gr.html (accessed Apr. 13, 2009).

[218] Carson, *Silent Spring*, 61.

insecticide residues, including DDT, are definitely not indestructible.

Of DDT in soil, Carson describes the process of earthworms eating soil and leaves, stating that "Undoubtedly some of the earthworms themselves succumb, but others survive to become 'biological magnifiers of the poison.'"[219] Then she describes robins eating the earthworms. She states, "As few as 11 large earthworms can transfer a lethal dose of DDT to a robin."[220] She goes on to report that an ornithologist at the University of Wisconsin, Joseph Hickey, reported in 1956 that 86 to 88 percent of robins died in sprayed versus unsprayed areas.[221] Hickey was a lifelong anti-DDT activist. Later we will report on his peregrine falcon work and his discredited participation in the 1971–72 EPA hearing against DDT.

A study on the Niagara Peninsula of Ontario, Canada, refutes the widely held belief in the continuing harm of DDT residues in soil and shows that robins are not harmed by high DDT residues in soil.[222] The study documented apple orchards as having been sprayed intensively with DDT through 1973. In fact, DDT use on orchards dates back to the very beginning of agricultural use of DDT. More DDT was used in the orchards than on any other commercial crop. Robins resident in the orchards were shown to have higher body burdens of DDT than predatory birds or any other bird, mammal, or invertebrate species. The tendency of robins to accumulate large body burdens of DDT from earthworms is why we stated earlier that of all birds, robins should have been the most harmed from DDT. Yet, they were not.

---

219 Ibid., 108.

220 Ibid.

221 Ibid., 109.

222 Harris M.L., et al., "Transfer of DDT and metabolites from fruit orchard soils to American robins (*Turdus migratorius*) twenty years after agricultural use of DDT in Canada," *Arch Environ Contam Toxicol*, vol. 39, no. 2 (2000): 205–20.

The Niagara Peninsula study compared orchards with high soil levels of DDT to surrounding areas with no DDT. There were no differences in earthworm abundance in soils with and without DDT. The high levels of DDT in the robins' food chain in orchards had no effects on reproduction. Numbers of robins were not affected by the presence or absence of DDT in soils. The study's authors concluded that aged DDT in soil is less toxic than the freshly sprayed chemical and that such changes in toxic actions of aged chemical had already been demonstrated for invertebrates. In fact, a 1998 study in New York reported losses in toxicity of DDT in sterile soil (heat treated to eliminate any living creatures, e.g., bacteria, etc.) in as little as thirty days.[223] Thus, because the soil was sterile, biodegradation was not possible. This is to say that bacteria and other life forms were not breaking DDT down into less complex chemicals. Vigorous chemical extraction showed that much of the DDT was still present, so loss of its toxic activity was a result of it being so tightly bound to organic matter that the DDT was not accessible to living organisms (or not bioavailable).

A follow-up study of the impact of DDT on robins in Canada was published in 2003. The investigators stipulated that "Although American robins nesting in Okanagan orchards are exposed to high OC [DDT] levels, reproductive success does not appear to be negatively impacted."[224] In fact, the authors found that numbers of eggs laid and numbers of baby birds were larger in orchards with DDT than in areas without DDT.[225]

[223] Robertson, BK., Alexander, M., "Sequestration of DDT and dieldrin in soil: Disappearance of acute toxicity but not the compounds." *Environ Toxicol Chem* 17 (1998):1034-1038.

[224] Gill, H., et al., "An assessment of DDT and other chlorinated compounds and the reproductive success of American robins (*Turdus migratonrius)* breeding in fruit orchards," *Ecotoxicology*, vol. 12, nos. 1–4 (2003): 113–123.

[225] Ibid.

These results stand as stark testimony to three facts: (1) DDT is not a major continuing source of wildlife harm even in the most highly contaminated soils, (2) DDT mostly stays where it is sprayed, and (3) DDT in soil quickly becomes harmless because it is either broken down or it becomes so tightly bound to soil as to not be available to living organisms.

The DDT–robin scare should be recognized and condemned for the lie it was. The public was misled and the entire story was fraudulent. Unfortunately, the DDT–robin scare does not end here.

Just as some scientists jumped on the bandwagon and advanced their careers through propagating misinformation about DDT, others were widely denounced for speaking the truth. Good and honorable scientists who spoke out against the DDT–robin-scare stories were publicly libeled and defamed. In one noteworthy case of defamation, professors Thomas Jukes, J. Gordon Edwards, Bob White Stevens, Donald Spencer, and Norman Borlaug (Nobel Peace Prize winner and recipient of a Congressional Gold Medal) were, on August 14, 1972, each described in the *New York Times* as "paid liar[s]."[226] The newspaper refused to print Jukes' rebuttal to the libelous charge, and some of the scientists sued. A jury trial was held in the United States District Court for the Southern District of New York, before Judge Charles M. Metzner. The jury decided in their favor and found that the *New York Times* and a member of the Audubon Society had defamed the claimants. Interestingly, Victor Yannacone, a lawyer who litigated against DDT in the

---

[226] 556 F.2d 113, 2 Media L. Rep. 1849, J. Gordon Edwards, Ph.D., Thomas H. Jukes, Ph.D., Robert H. White-Stevens, Ph.D., Plaintiffs-Appellees, v. National Audubon Society, Inc., Robert S. Arbib, Jr., Elvin J. Stahr, Jr., Defendants, and The New York Times Company and Roland C. Clement, Defendants-Appellants. Nos. 1026, 1027, Dockets 77-7018, 77-7034. United States Court of Appeals, Second Circuit. Argued May 10, 1977. Decided May 25, 1977. http://bulk.resource.org/courts.gov/c/F2/556/556.F2d.113.77-7034.77-7018.1026.1027.html. (accessed Sept. 24, 2009).

1960s, testified in the case that the defendant for the Audubon Society was present at an Environmental Defense Fund (EDF) meeting when the use of smear tactics in the press against those who spoke in favor of DDT was discussed.[227]

On appeal to the United States Court of Appeals of the Second Circuit, Judge Irving Kaufman opined that the claimants had indeed been defamed without "any serious basis":

> "To call the appellees, all of whom were university professors, paid liars clearly involves defamation that far exceeds the bounds of the prior controversy. No allegation could be better calculated to ruin an academic reputation. And, to say a scientist is paid to lie implies corruption, and not merely a poor opinion of his scientific integrity. Such a statement requires a factual basis, and no one contends there was any serious basis for such a statement in this case.[228]

Nevertheless, Kaufman vacated the lower court's verdict, justifying his decision as follows:

> "It is unfortunate that the exercise of liberties so precious as freedom of speech and of the press may sometimes do harm that the state is powerless to recompense: but this is the price that must be paid for the blessing of a democratic way of life."[229]

In general, the basis for this ruling seems to be the *New York Times* was not found liable because it accepted the report from a reputable organization (The National Audubon Society) and the

---

[227] Footnote 6: http://bulk.resource.org/courts.gov/c/F2/556/556.F2d.113.77-7034.77-7018.1026.1027.html.

[228] Footnote 5: http://bulk.resource.org/courts.gov/c/F2/556/556.F2d.113.77-7034.77-7018.1026.1027.html.

[229] Section V, paragraph 58: http://bulk.resource.org/courts.gov/c/F2/556/556.F2d.113.77-7034.77-7018.1026.1027.html.

Audubon Society was not liable because of insufficient evidence of malicious intent.

There was no justice for the defamed professors in the courts. Yet, they were men of great integrity and courage and they deserve our respect. They stood against a stampede of public opinion and politics. They spoke the truth even as the public was being assaulted by an ever-growing wave of misrepresentations and falsehoods.

The Robin Story: Lessons Learned

Today in the United States, each of us can look outside in the spring and know with certainty that robins are abundant and healthy. There are even suggestions that the robin's continuing increase in numbers might be detrimental to other birds and that management strategies might be needed.[230] Yet, this bird has a special place in people's hearts and minds. Few can see a robin hopping about the yard in spring, gathering nesting twigs or tugging to dislodge an earthworm, without a sense of joy and wonder at the renewal of life each spring and the marvelous beauty and complexity of the living world. It was this special relationship between birds and humans that Carson targeted in her campaign of fear.

Carson's robin story riled the public. No doubt it was also helpful in opening the editorial doors of prestigious journals to all the anti-DDT claims yet to come. However, under attack for false claims of robin endangerment, insecticide opponents had to change their claims about DDT as a cause of robin extinction. This is illustrated in an exchange in *Science* magazine between Thomas Jukes and Charles Wurster, a founder of the Environmental Defense Fund (EDF), an environmental public-interest law firm. Wurster, as we explain in more detail below, was pivotal in the fight against DDT and in his response to a

[230] DeAnna, P. "The American robin: American success story?"

letter by Jukes, Wurster stated, "Jukes has erected, then demolished, the abominable straw man of robin extinction. We never mentioned this point however, nor have other scientists."[231] For certain the robin–DDT story provided a lesson to activists within the anti-insecticide movement; *viz.*, that an already endangered species would need to be used to advance their campaign against DDT.

It is our contention that the basic adjustment in the anti-DDT movement's strategy was to focus only on species that were already threatened so at least a claim of possible extinction would be beyond refute. However, the change in strategy was not without problems. One such problem occurred when Wurster published a claim, again in *Science*, that DDT threatened the Bermuda petrel bird with extinction.[232] He suggested in 1968 that DDT could cause extinction of the petrel by the mid-1970s. But there was no definable link between an ocean-going (pelagic) bird and potentially harmful levels of DDT. To our knowledge, no papers reporting to confirm his observations were ever published.

Wurster's meritless claim brought an immediate response. Louis McLean, consultant to the Industry Task Force for the defense of DDT in the late 1960s,[233] wrote to the editors of the magazine to protest. His letter, published July 23, 1968, said,

> Wurster and Wingate . . . state that Stickel . . . found DDT residues of 10.6 ppm in bald eagle eggs, that others have reported DDT residues of 120 to 227 ppm in herring gull eggs, and that both species are declining.

---

[231] Wurster, C.F., "DDT and robins," *Science*, vol. 158, no. 822 (1968): 1413–14.

[232] Wurster, C.F. and Wingate, D.B., "DDT residues and declining reproduction in the Bermuda Petrel," *Science*, vol. 159 (1968): 979–81

[233] Moore, W.G., "The Wisconsin Ban on DDT: Old Law, New Content," http://www.law.wisc.edu/alumni/gargoyle/archive/16_2/gargoyle_16_2_2.pdf. (accessed Sept. 24, 2009).

> From this they reason that the suggestion is "that susceptibility varies widely between species," and they conclude that residues of 6.44 ppm in the petrel eggs may account for the fluctuating decline in the reproductive success of the small, protected, and increasing population of the once-considered-extinct Bermuda petrel.
>
> With regard to Stickel's information . . . he states that the residues in the eagle eggs "provide little basis for suspecting that DDT in the eggs prevented hatching." Stickel further notes that, while an "occasional" eagle may obtain enough pesticide to place it "in hazard," "most eagles that die in the United States today die of causes other than pesticide poisoning."
>
> Herring gulls and ring-billed gulls have undergone a population explosion to such an extent that the U.S. Fish and Wildlife Service . . . has considered measures to reduce their numbers. Dr. W. H. Drury . . . director of research, Massachusetts Audubon Society, has urged the immediate burying of garbage at dumps to stop the population explosion of the gulls.
>
> Even if the fluctuations in the ability to hatch of the eggs of the few pairs of Bermuda petrels observed by Wurster and Wingate are not within normal limits for that little-known species, Stickel's comments and gull populations provide little basis for suggesting that DDT residues of 6.4 ppm, which is less than the amount permitted in foods, are of significance."[234]

History has shown that McLean was right and Wurster was wrong. McLean exposed the falsehoods of Wurster's claims that DDT was endangering the petrel and that herring gulls were declining. Herring gulls were rapidly increasing in numbers and the Bermuda petrel was not threatened at all by DDT. Much is

[234] McLean, L.A., et al., "DDT residues and Bermuda petrels," *Science* July 23, 1968, 397.

now known about the Bermuda petrel,[235] and additional information is presented in appendix 2.

With the claim of robin extinction proven wrong and the Bermuda petrel story a bust, DDT opponents still needed a bird species that was truly threatened, one that they could then claim was threatened because of DDT contamination. With time, the bald eagle story became the public-relations success that environmentalists needed. The wide acceptance and popularity of *Silent Spring*, where Carson had briefly described the bald eagle as being harmed by DDT, prepared the public and scientists to readily accept other claims about DDT and eagles. But while Carson was the one to put this story before the public, it fell to others to shape the story into a media event and an international phenomenon.

The Bald Eagle Story

To transition from the robin to her main target, the bald eagle, Carson wrote, "Like the robin, another American bird seems to be on the verge of extinction."[236] Carson claimed that DDT was central to the decline in bald eagle populations, and she reported on findings by a retired banker, Charles Broley, who monitored eagle populations in Florida.

> During the early years of his banding, Mr Broley used to find 125 active nests a year on the stretch of coast he had chosen for his work. The number of young banded each year was about 150. In 1947 the production of

[235] Bird Life International, "Bird life species factsheet—Bermuda petrel. Additional data," http://www.birdlife.org/datazone/species/index.html?action=SpcHTMDetails.asp&sid=3910&m=1 (accessed Apr. 13, 2009); National Audubon Society, "Birds and science—watch list. Bermuda petrel," http://web1.audubon.org/science/species/watchlist/profile.php?speciesCode=berpet (accessed Apr. 13, 2009).

[236] Carson, *Silent Spring*, 118.

> young birds began to decline. Some nests contained no eggs; others contained eggs that failed to hatch. Between 1952 and 1957, about 80 percent of the nests failed to produce young. In the last year of this period only 43 nests were occupied. Seven of them produced young (8 eaglets); 23 contained eggs that failed to hatch; 13 were used merely as feeding stations and contained no eggs. In 1958 Mr Broley ranged over 100 miles of coast before finding and banding one eaglet. Adult eagles, which had been seen at 43 nests in 1957, were so scarce that he observed them only at 10 nests.[237]

To Carson, the central cause of this observed decline in bald eagles was clear: "In seeking an explanation, only one appears to fit all the facts. This is that the reproductive capacity of the birds has been so lowered by some environmental agent that there are now almost no annual additions of young to maintain the race."[238]

The environmental agent that Carson identified was, of course, DDT. She stated that only one explanation fit all the facts; it was, of course, the only explanation she wanted. Carson presented no data or observations on other possibilities, or even mentioned other possible explanations for declining numbers of bald eagles. She noted that DDT had been used widely in Florida and stated, "the eagles have almost certainly been storing up the DDT in the tissues of their bodies. And . . . they are less and less able to produce young and to preserve the continuity of their race."[239]

In the decades that followed the publication of *Silent Spring*, environmentalists and scientists perpetuated the idea that DDT not only devastated bald eagle populations, but the subsequent

---

237 Carson, R., *Silent Spring* (Aylesbury, Bucks, UK, Penguin Books: 1972), 113.

238 Ibid., 115.

239 Ibid., 116.

banning of DDT in agriculture was directly responsible for their recovery. These claims against DDT are false. Environmentalists and scientists also perpetuated Carson's practice of ignoring other, far-more important explanations for wildlife harm. Indeed, Carson and her followers might have been well served by going back to a classic text on North American wildlife, *Hornaday's American Natural History*. William T. Hornaday was an internationally recognized conservationist, and, at one time, the director of the New York Zoological Park. In the 1935 edition of his book, he wrote about the bald eagle:

> To-day, throughout the whole United States the American Eagle [bald eagle] is a scarce bird, and growing more so, day by day. You now can travel the whole length and breadth of Florida without seeing even one; which I can testify means a vast falling off since 1900. Old Baldy is passing out, along with the other birds, and it is probable that millions of Americans now living never will see one alive.
>
> In 1904, live eagles were easily purchasable; now they are difficult to obtain, and costly.

Hornaday's assessment of what was happening to bald eagles was authoritative and non-ideological. Long before DDT was ever used, he declared that bald eagle populations were declining at rapid rates and that bald eagles were already endangered in Florida and the rest of the United States.

Decline and Recovery of Wildlife Populations

The bald eagle story is central to many campaigns against DDT, including the campaign against its public-health use.[240] In

[240] In a discussion of DDT's role in public health, Henk van den Berg draws attention to the environmental effects of DDT, stating that "studies have shown that DDT is highly toxic to insects, shrimps and

order to understand the reasons for the decline and rise in bald eagle populations, we will briefly review trends of wildlife populations in the United States during decades leading up to the 1960s. We will then focus on the 1960s because this was the cataclysmic period, the decade when public attention was falsely directed at DDT as a mega-threat to wildlife and the environment.

In the decades leading up to the 1960s, wildlife populations throughout the United States endured severe and devastating harm—but not from insecticides. As we will show in coverage of different wildlife species, harm was due to unregulated hunting. Throughout human history hunting had been a central part of human existence and experience. Hunted meat was an important part of the early settlers' diet in the United States. Settlers lived off the land and harvested game not only for food, but also for commerce and sport. Human population pressures and the need for meat to feed a youthful and rapidly growing population influenced the pace of unregulated hunting. As revealed in our coverage of endangered species, below, unregulated hunting continued to be a problem for species survival through much of the twentieth century.

The harm this caused to wildlife populations is best illustrated by the story of the North American bison.[241] Through such events as the arrival of the horse on the Great Plains in the early 1700s, various Native American treaties, epidemics, and the beginning of the robe market in 1820s, the bison was extinct east of the Mississippi by the 1820s. The destruction of bison populations west of the Mississippi continued with the

---

fish and adversely affects the reproduction of wild birds through thinning of egg shells. Van den Berg, H., "Global status of DDT and its alternatives for use in vector control to prevent disease," *Environ Health Persp* (2009), doi: 10.1289/ehp.0900785, p. 14.

[241] The American Bison Society, section entitled "Bison Timeline" and section on "History of ABS." http://www.americanbisonsocietyonline.org/Home/tabid/169/language/en-US/Default.aspx. (accessed Sept. 24, 2009).

development of a more commercial market for buffalo hides in 1871, penetration of railroads into the Great Plains, and the development of Sharps .50-caliber rifle in 1872. By 1889, Hornaday found just 1,091 American bison remaining in North America. He went on to establish the American Bison Society (ABS) in 1905, an organization that was ultimately helpful in preventing the total extinction of bison from the continent.

Initially, settlers and hunters systematically harvested predators because they competed with humans for food and because their fur, feathers, and pelts were valuable commodities. While national reliance on hunted meat began to decline as people transitioned from rural to urban living and began to rely on locally produced farm commodities, hunting remained, and continues to be, a major American pastime.

The development of Sharps .50-caliber rifle in 1872 that helped speed the crash of American bison populations is just one example of how advances in the power, velocity, and accuracy of rifles and shotguns influenced the future of North American wildlife.[242] Another notable improvement was the hunting scope, which started becoming more widely available in the 1920s.[243]

During WWII, however, the war temporarily stopped the killing of many forms of wildlife in North America. Suddenly, the most aggressively sought species were freed of persecution

---

[242] "Continued evolution of hunting tools resulted in the development of the crossbow with its heavier and shorter quarrels or bolts. Matchlock, wheel lock and flintlock firearms followed these rather quickly. Flintlocks remained the dominant type of hunting tool for a couple of centuries before being replaced by the caplock, which led rather quickly to the development of modern firearms. The time between innovations became progressively shorter and continuing changes in earlier technologies [are] still taking place." Gun Muse Outdoors, "History of hunting," http://www.gunmuse.com/Main/Firearm+History/4-H+Program+Guide/Philosophy+Laws+and+More/History%20of%20Hunting (accessed Apr. 13, 2009).

[243] Optics Planet.com, "Docter Optics," http://www.opticsplanet.net/docter-brand.html (accessed Apr. 13, 2009).

because of the enormous movement of people out of North America in order to fight the war. The severe rationing of gasoline for transportation, the prohibition of luxury driving, the hoarding of metal (copper was needed for munitions and was not even available for minting of pennies[244]), and the huge need for ammunition to support the war effort also impacted hunting. As reported on Monday, July 1, 1940, on the day that new tax levies were imposed by the emergency defense tax bill, "Few, if any, will escape without contributing their share—large or small—of the tremendous cost of building up the national defenses."[245] And, as we now know, none escaped, to include hunters and hunting.

However, the temporary respite afforded hunted wildlife in the United States by conduct of WWII ended with the ending of the war. Along with firearm improvements that had occurred during the war, young citizens had been drafted or volunteered for service, put in uniforms, given firearms, and taught to shoot. After the war, they came home and used their familiarity with firearms and their marksmanship to hunt wildlife. This had an impact on the bald eagle and peregrine falcon populations. In fact, membership in the National Rifle Association (NRA), created by members of the retired military in 1871 to improve marksmanship,[246] tripled immediately after WWII. Furthermore, as a book on the history of the organization by an anti-NRA

---

[244] uspatterns.com. J2081/P2077: http://www.uspatterns.com/p2077.html (accessed August 16, 2009).

[245] "Arms taxes go in effect." The Daily Review, Clifton Forge, Virginia, Monday, July 1, 1940, 1 and 6.

[246] "Dismayed by the lack of marksmanship shown by their troops, Union veterans Col. William C. Church and Gen. George Wingate formed the National Rifle Association in 1871. The primary goal of the association would be to 'promote and encourage rifle shooting on a scientific basis,' according to a magazine editorial written by Church." National Rifle Association, "A brief history of the NRA," http://www.nrahq.org/history.asp (accessed Apr. 13, 2009).

advocate, Josh Sugarmann, explains, the NRA's emphasis changed from marksmanship to hunting:

> Starting in 1946 the organization finally began to expand beyond its historic base of military and civilian target shooters to include hunters. This new focus rose in part from the large number of World War II veterans who, having been exposed to firearms, had taken up hunting upon their return home. Shooting etiquette learned on the battlefield was, however, viewed less fondly on the home front by the general public. As a result, the NRA instituted a hunter training program and began lobbying for mandatory hunter safety training in states.[247]

This change in emphasis has also been recognized by other organizations. Another anti-NRA advocacy group, Sane Guns, writes, "The first century of the NRA's history (from it's [*sic*] incorporation in 1871 to the 'Cincinnati Revolt' of 1977) was directed towards promoting shooting sports—first marksmanship and then (following WW II) hunting."[248]

Changes in land use and pressure on the natural environment were other sources of wildlife harm. Growing densities of human populations and the clearing of land for agriculture, which was accelerated by the advent of modern equipment, changed the landscape and eliminated natural habitats. Habitat pressures were greatest in areas densely populated by humans and by and large, those densities were greatest in the East and least in the West. This aspect of U.S. demography is revealed in statistics on the east-west distribution of people in the United States. The center for average U.S. population from 1790 to

---

[247] Sugarmann, J., *NRA: money, firepower & fear* (Palo Alto: National Press Books, 1992), http://www.vpc.org/nrainfo/chapter1.html (accessed Apr. 13, 2009).

[248] Sane Guns, "The gun lobby. The NRA," http://www.saneguns.org/gunlobby/nra.html (accessed Apr. 13, 2009).

2000 has gradually shifted from east to west. In 1800, the center was in Maryland; in 1900, it was in Indiana; and in 2000 it was in Missouri.[249] All in all, the late nineteenth century and first half of the twentieth century were times of extensive habitat destruction and unregulated private, commercial, and sport hunting of all forms of wildlife.[250]

The Recent History of Bald Eagles in the United States

Bald eagles that nest in northern regions are referred to as northern bald eagles. Likewise, those that nest in southern areas are referred to as southern bald eagles. Northern bald eagles migrate to southern zones in winter, and southern birds migrate to more northern zones in the summer. Additionally, some bald eagles are classified as permanent residents of an area. Even permanent residents, however, wander away from nesting sites

---

[249] U.S. Census, "Mean center of the population for the United States: 1790 to 2000," http://www.census.gov/geo/www/cenpop/meanctr.pdf (accessed Apr. 13, 2009).

[250] "Unregulated hunting has at various times threatened the existence of some game animals. Prime examples of this occurred in the United States in the 19th century, when egrets were decimated for their plumes, used in millinery. Buffalo herds (*see* Bison) that once numbered in the millions were also virtually wiped out by hunters intent on profit and enjoyment; by 1895 only 400 buffalo remained in the country. Hunters took their toll on other wildlife as well. Pronghorn, deer, and mountain sheep were decimated to feed the growing population in the West. In the East, waterfowl populations plummeted as hunters armed with cannonlike [*sic*] guns wiped out entire flocks in one shot and then shipped the birds to restaurants to satisfy exotic tastes. Some game birds, such as the passenger pigeon and the heath hen, were hunted into extinction by people who believed the game supply to be infinite." Encarta Encyclopedia, "Hunting," http://encarta.msn.com/encyclopedia_761576243/Hunting.html (accessed Apr. 13, 2009).

at different times. For all these reasons, the bald eagle is characterized as a partial-migratory bird.

Bald eagles are long lived, sometimes living for thirty years or more in the wild and even longer in captivity. They mate for life, and both males and females are involved in nest building and the rearing of offspring. Newly hatched eagles go through a learning process about their surroundings. Year after year they will return to the same nesting area and build their nest there. Eagles lay one to three eggs a year, but eaglets compete with siblings for food, so depending on food availability, only a single eaglet may survive. Bald eagles are considered juvenile for a period of about five years, until their head feathers become white, giving them the stunning white head that characterizes the species. As juveniles, they are similar in appearance to golden eagles.

Eagles are predators and often feed on fish, but they also feed on other birds and animals. However they also will scavenge and the scavenging lifestyle makes them susceptible to poisoning, should people deliberately leave poisoned carcasses out for eagles to consume.

The Skagit River Bald Eagle Awareness Team, a nonprofit group in Washington dedicated to bald eagle conservation, describes the history of bald eagle population decline in the United States.

> As many as 500,000 bald eagles once soared across North American skies. The last 300 years, though, has seen a huge decline in North America's bald eagle population. Early colonists feared eagles could kill livestock or even carry away sheep. Many eagles were killed out of fear—or for sport. It started early. A report from the year 1668 recorded that many of an "infinite number" of eagles wintering in Casco Bay, Maine, were shot and fed to hogs. The bald eagle was declared the U.S. national symbol in 1782 but that did little to stop the carnage. Between 1917 and 1953, the Alaskan

> legislature endorsed bounties on bald eagles—seen as a threat to salmon runs—resulting in over 128,000 being killed. Even more recently, Wyoming ranchers paid helicopter pilots and rifleman 25 dollars for each eagle they killed.[251]

Data from the Missouri Department of Conservation illustrate the major trends of bald eagle populations in Missouri:

> Nesting bald eagles were common in Missouri in the early 1800s. By 1890, they were nearly eliminated as nesters. Before the cypress forests were cut and the swamps drained, the main concentration of nesting bald eagles was in the cypress swamps of the Bootheel . . . Missouri's eagles were already gone by the mid-1900s.[252]

Other states report similar histories of bald eagle population decline. Nesting had ceased in Oklahoma.[253] As described by two Oklahoma wildlife biologists, James Lish and Steve Sherrod,

> If any nesting Bald Eagles survived in Oklahoma until the end of Indian occupation (which ended shortly after the Civil War) it is doubtful that many could have survived the biological disasters that were to come at the

---

[251] Skagit River Bald Eagle Awareness Team, "Interesting facts about bald eagles," http://www.skagiteagle.org/Wildlife/eagle-facts.htm (accessed Apr. 13, 2009).

[252] Missouri Department of Conservation, "The bald eagle in Missouri," http://mdc4.mdc.mo.gov/Documents/12138.pdf (accessed Apr. 13, 2009).

[253] Lish, J.W. and Sherrod, S.K., "A history of bald eagle nesting activity in Oklahoma," *Proc Okla Acad Sci*, vol. 66 (1986): 15–20, http://digital.library.okstate.edu/oas/oas_pdf/v66/p15_20.pdf (accessed Sept. 25, 2009).

> turn of the century. These were the land rush, the widespread deforestation that subsequently occurred, and the well documented poisoning, trapping, and general extermination of predators by settlers. Few, if any other, states have had such widespread ecological destruction occur over such a short period of time. In 1912 Barde . . . wrote that the opening of Oklahoma "to homestead settlement was as though the lightenings of heaven had fallen upon its wild life. Fires swept the country in all directions; armies of horsemen penetrated to remote places; wagons rumbled over the hills and down the valleys; there was a tumult of shouting and the roar of guns, and violence reigned, where until that day had been the peace of a primeval solitude. The slaughter had begun." It is no surprise therefore that little evidence of Oklahoma's original nesting Bald Eagle population has been preserved.[254]

Perhaps conditions were somewhat atypical in Oklahoma at the turn of the century, but elimination of nesting bald eagles occurred, albeit at different times, throughout the United States:

- 1949 witnessed the last confirmed attempt of eagles to nest in Alabama prior to the recovery efforts of the 1980s.[255]
- In 1974, there was only one pair of nesting bald eagles in Colorado.[256]

[254] Ibid. James W. Lish was with the Oklahoma Cooperative Wildlife and Fisheries Research Unit, Oklahoma State University, Stillwater, Oklahoma, 74078; Dr. Steve Sherrod was head of the George Miksch Sutton Avian Research Center in Bartlesville, Oklahoma, in 1990.

[255] Outdoor Alabama, "Bald Eagle," http://www.outdooralabama.com/watchable-wildlife/birding-wildlife/bald-eagles.cfm (accessed Apr. 13, 2009).

- South Dakota reports that "although historically bald eagles nested in southeastern South Dakota, there have been no confirmed nesting records in the state" during the twentieth century.[257]
- The Northern States Bald Eagle Recovery Plan states that eagles in Indiana were "no longer known to breed east and south of the Wabash River" in 1897.[258]
- Eagles were thought to have been common in Kansas, but there are "no verified records of successful nesting in the 20th Century."[259]
- Eagles were once very common in Maine, but "by 1908 the number of pairs was estimated at not over 100," and eagles were "considered very rare."[260]
- The eagle was "formerly a common breeder" in areas of Nebraska, but "no verified record exists of successful nesting in the 1900's."[261]
- In New York, bald eagles were "commonly observed until nesting and wintering populations began to decline in the early 1900's."[262] In the

---

256 Roaring Fork Conservancy, "Bald Eagle *haliaetus leucocophalus*," http://www.roaringfork.org/images/other/Bald_Eagles.pdf (accessed Apr. 10, 2009).

257 Northern State University, "Bald Eagle (haliaeetus leucocephalus)," http://www.northern.edu/natsource/ENDANG1/Baldeg1.htm (accessed Apr. 13, 2009).

258 Northern States Bald Eagle Recovery Team, "Northern states bald eagle recovery plan," U.S. Fish and Wildlife Service, July 29, 1983, p. 13, http://www.fws.gov/Midwest/eagle/recovery/recovery.html. (accessed Sept. 25, 2009).

259 Ibid., 14.

260 Ibid.

261 Ibid., 16.

262 Ibid., 17.

1960s, only one pair of nesting eagles was present in New York.[263]

- Iowa reports that "direct persecution (mostly shooting) and changes in eagle habitat, particularly nesting habitat, appear to have eliminated the bald eagle as an Iowa nester by the early 1900s" and that "eagles were 'formerly common in Iowa and frequently nested in favorable localities.'"[264]

The elimination of bald eagles occurred progressively over many decades, but accelerated in the 1800s and first half of the 1900s. The population declines corresponded with growth in human populations and increasing habitat destruction. By the time DDT came into wide use in 1946, nesting populations of bald eagles were already in danger of extinction in the United States, and only a few relic populations of nesting eagles remained.

The Eagle Protection Act of 1940 was passed as an early attempt to avert extinction. The act addressed the single greatest threat propelling them toward that outcome by specifically outlawing the hunting and killing of bald eagles (referred to as "taking"). Relic populations of nesting eagles that did persevere did so because of their remote or protected locations, unusually favorable ecological conditions, or just good fortune. But human-development pressures continued to grow everywhere, and as wealth increased and hunting technology improved, pressures against eagle survival grew ever stronger. Even if nesting sites were in adequately protected locations, the eagles roamed widely, which meant that they flew into ever-more dangerous skies.

[263] Roaring Fork Conservancy, "Bald Eagle *haliaetus leucocophalus*,"

[264] Iowa Department of Natural Resources, "Bald eagle restoration," http://www.iowadnr.com/wildlife/pdfs/eagle%20log%202003%20complete.pdf (accessed Mar. 13, 2009).

In terms of population dynamics, eagles' risks were magnified because they mate for life. Thus, the tragic loss of a mate might not be immediately resolved by recruiting another mate. Furthermore, because eagles imprint on the nest sites where they fledge and return to those sites as adults, once nesting populations are eliminated from an area, the only way to rapidly get nesting populations back is to raise and fledge eaglets at those locations. The Eagle Protection Act of 1940 was designed to stop wholesale taking of eagles, but it did nothing to return nesting populations to areas from which they had been eliminated. The recovery phase would not come for more than thirty years.

DDT: A Threat to Bald Eagles?

The bald eagle is the national symbol of the United States, and in 1962, Carson portrayed it as being under immediate threat of extinction by DDT. Her strategy was brilliant: A claim of threat to the eagle mobilized the public, politicians, and opportunists in the scientific community even though she presented no evidence that she had given any thought whatsoever to alternative explanations for the birds' plight.

Her dominant assertion was that DDT reduced bald eagle reproduction, and that this was the cause of reductions in numbers of bald eagles in the United States. In *Silent Spring*, she documents cases of reduced eagle populations and asserts that DDT was sprayed in areas where eagle populations were once abundant. The book quickly brought a possible impact of DDT to the public's attention, but there was no clearly defined mechanism for how DDT reduced bald eagle reproduction.

A core tenet of the early claim was that eagles had started to recover after the Bald Eagle Protection Act of 1940, but that wide use of DDT stopped the recovery and sent eagle populations into steep decline. However, the Bald Eagle Protection Act was designed to curtail hunting of bald eagles and was passed before DDT was ever used. Clearly, back then legislators and wildlife professionals knew exactly what the real threat to bald eagles was: hunting. Furthermore, the Bald Eagle Protection Act did not really stop the taking of eagles. Perhaps there was some lessening of the slaughter, but the law was not enough. This truth is confirmed by the documented history of bald eagles in Wisconsin, which states that "with the passage of the Migratory Bird Treaty in 1916 and the Bald Eagle Protection Act in 1940, it became illegal to shoot bald eagles. But enforcement of the law was weak. By 1950, eagles no longer existed in the southern two-thirds of Wisconsin."[265]

None of this information is presented as a condemnation of hunting or hunters. Indeed, with the elimination of natural predators, the need for hunting and hunters as a check on burgeoning game populations has never been greater than it is today. Most hunters are law-abiding citizens, and more importantly, most hunters cherish, protect, and respect wildlife. In fact, hunters have made an enormous contribution to the conservation of wildlife.[266]

Though the slaughter of bald eagles did not end after the passage of the 1940 act, it did decline. The claim that eagle populations were actually beginning to recover after passage of the act, however, is not supported by any meaningful collection of data, nor is there support for the alternative claim of recovery after the 1916 Migratory Bird Treaty Act. In fact, the extraordinary claims of eagle recovery by anti-DDT activists bring into question the meaning of the word "recovery" in context of wildlife management.

The white-tailed deer and the wild turkey are two other species that were in many areas broadly eliminated by human activities such as hunting and changes in land use. The fortunes

---

[265] Wisconsin Department of Natural Resources, "Bald Eagle *haliaeetus leucocephalus*, history in Wisconsin" http://ua.dnr.wi.gov/org/land/er/factsheets/birds/eagle.htm (accessed Apr. 13, 2009).

[266] GunMuse Outdoors, "History of hunting."

of these two populations in the United States illustrate the population dynamics of endangered species. They provide useful examples of the roles, both destructive and constructive, that man has played in wildlife management, and their stories highlight the fundamental management principles that promote population recoveries. We will use these principles to show what actually happened to bald eagles in the United States.

White-Tailed Deer and Wild Turkey in the United States

Today, most people, even in urban areas, consider white-tailed deer to be overly abundant, a danger on our highways, and a threat to our gardens. But it was not always so, as the history of the deer in Missouri illustrates. In pre-settlement times, deer were plentiful and distributed throughout the state.[267] In the early 19th Century there were as many as 700,000 deer, but with extensive settlement and farming, deer habitat was eliminated, and hunting further reduced deer numbers. Deer had disappeared from many counties in Missouri by the late 1800s, and by the early 1900s, deer had essentially disappeared altogether from Missouri.

The low point occurred in 1925 when the Game and Fish Department reported only 395 deer in 23 counties. Officials acted quickly to ensure the survival of relic populations. Deer hunting was closed that same year, and land was purchased for wildlife refuges and deer were imported to the refuges. This emphasis on better management helped, but there was no dramatic increase in deer. Slow population growth occurred because deer were present in too few locations. A statewide inventory in 1934–35 found only 2,240 deer in 28 southeastern Missouri counties. A comprehensive deer-restoration program

[267] Missouri Department of Conservation, "White tailed deer - *Odocoileus virginianus*," http://www.mdc.mo.gov/nathis/mammals/deer2/ (accessed Apr. 15, 2009).

was initiated in 1938 that emphasized introducing deer to receptive habitats throughout the state. By 1944, there were more than 15,000 deer. Deer restocking continued until 1957 when deer were present in all counties. The deer population continued to grow rapidly in the 1960s, and current estimates put the herd at over 750,000 animals.

Deer elimination occurred in Missouri and most other states because there were no hunting restrictions, and there was a high demand for venison. Also, deer were shot for sport and for trophies. The destruction of deer habitats further contributed to their decline. Forest habitats were extensively eliminated through heavy logging and the clearing of land for development and agriculture. Deer flourish in forest habitats or thickets, so even if they had not been eliminated through widespread hunting, their populations would have declined along with the destruction of their natural habitats.

The deer were nearly gone by the early twentieth century, but with Americans' growing affluence and increasing leisure time, the public demanded return of this popular game animal. Public support, appropriate legislation, public policy, and good management strategies converged to allow a full recovery of the white-tailed deer. Deer recovery also demonstrates how people were willing to work together through decades of struggle in order to restore deer.

Deer received priority attention because they were the most popular game animal in the United States. Yet, the whole spectrum of wildlife species in North America had been over hunted, not just white-tailed deer. The extinctions of the passenger pigeon and heath hen are tragic examples.

Putting habitat destruction and unregulated hunting aside, there are important lessons to be learned from the recovery of deer in Missouri. Achieving maximum growth over a large area like the state of Missouri requires as many sites of rapidly reproducing deer as possible. This was, in effect, what happened. Now, decades later, deer are as abundant and widely distributed as they were originally. Through restocking,

Missouri officials tapped into a basic dynamic of wildlife management. Deer introduced into a new area with abundant food and appropriate cover and protections (e.g., forests, thickets, and the edges of forests or thickets[268]) can reproduce at maximum rates.

The deer story is not unusual. The wild turkey has a similar history and experience, as the Missouri Department of Conservation describes:

> Missouri turkey population had been reduced from 250,000 to approximately 3,000 birds and most were limited to remote Ozark regions. With the destruction of timber by fire and ax, open grazing of the forest, and market hunting, the wild turkey in Missouri was brought to near extinction.
>
> From 1925 to 1943, an effort was made to halt the decline in turkey numbers. Approximately 14,000 game farm turkeys were released to supply more breeding stock. The hunting season for turkeys was closed in 1937. However, turkey numbers still decreased and the release of game farm turkeys appeared futile. Research studies were initiated in 1938 to determine why the game farm turkeys failed. These investigations were interrupted during World War II, but resumed in 1951. It was learned that game farm birds, even of the wildest stock obtainable, were not the answer—only a truly wild bird would survive.

[268] University of Michigan, Museum of Zoology, "*Odocoileus virginianus* - white-tailed deer," http://animaldiversity.ummz.umich.edu/site/accounts/information/Odocoileus_virginianus.html (accessed Apr. 15, 2009).

> It became evident that restocking wild turkeys to unoccupied range was the solution to restoring the species."[269]

In the end, the department of conservation bought 23,000 acres of land in the Ozarks that still supported a small relic population of wild turkeys. Good research and management practices allowed the service to begin capturing and restocking from that site in 1957. Turkeys have higher reproduction rates than deer, and recovery was rapid. After twenty-three years of no turkey hunting, hunting was again opened in April 1960 for just three days in only fourteen counties. By 1985, the conservation service had open seasons in all counties, and in 2000, hunters harvested over 70,000 turkeys in Missouri.

From our description of the recovery of wild turkeys in Missouri, it seems certain that wildlife professionals would not have characterized success in propagating a relic population (an isolated surviving population) of wild turkeys in a remote area of Missouri's Ozark region as a "recovery" of wild turkeys. Recovery occurred when turkeys were taken from the Ozark site and successfully introduced into areas from which they had been eliminated. Through those introductions, wildlife managers and the public leveraged the reproductive potential of turkeys by placing them in as many favorable locations as possible. Those introductions, along with legal protections of introduced populations, brought about "recovery" of wild turkeys in Missouri.

As the examples of deer and wild turkeys demonstrate, a consistent management strategy for wildlife recovery is to get endangered species back into as many areas as possible, as quickly as possible. There are multiple reasons such a strategy is critical. One purpose of widely distributing populations is to

[269] Missouri Department of Conservation, "The wild turkey in Missouri," http://mdc.mo.gov/nathis/birds/turkey/index.htm. (accessed Sept. 25, 2009).

avoid the possibility of a natural disaster causing extinction by eliminating just one or two relic populations. Such disasters can occur anywhere and can spring from many sources, including severe weather events, unusual levels of natural predation, and epidemics of natural diseases—or any combination thereof. Such a convergence of unfortunate events contributed to extinction of the heath hen. Christopher Cokinos, poet and nature writer, has described the many factors that led to heath hen extinction.[270] His writings were confirmed in an earlier published book by the Massachusetts state ornithologist, Edward Howe Forbush of the early 20th Century.[271] In a separate review, the circumstances of heath hen extinction have been described as follows:

> The last birds were wiped out by a series of relatively common, but deadly, natural events: fire, starvation, predation, and disease. But the heath hen's continued existence as a species would not have been so vulnerable to these occurrences if their populations had not been severely reduced already by human hunting. In its former range, the heath hen easily could have survived any one of these stresses, or even all of them in combination.[272]

From these analyses, we can conclude that the heath hen was not brought to extinction solely by natural events: excessive hunting pressure was the primary cause. So how do the stories of

---

[270] Cokinos, C. *Hope is the thing with feathers: A personal chronicle of vanished birds.* (New York: P.P Tarcher, 2001) (chapter on heath hen begins on pages 121).

[271] Forbush, E.H. *A history of the game birds, wild-fowl, and shore birds of Massachusetts and adjacent states.* (Boston: Massachusetts State Board of Agriculture, 1912)

[272] Bagheera: In the Wild, "Extinct in the wild: heath hen," http://www.bagheera.com/inthewild/ext_heathhen.htm (accessed Apr. 13, 2009).

the white-tailed deer, the wild turkey, and the heath hen relate to the threat of extinction of bald eagles by DDT?

Protecting small relic populations in a few areas of the United States with the Eagle Protection Act did not constitute recovery of nesting populations of bald eagles in the United States. Regardless, the environmentalists' claim that eagles were recovering prior to wide use of DDT is an important claim. It is particularly important because anti-DDT groups used the claim of recovery to justify their campaigns, and more important, to claim that DDT was the primary force behind the threat of eagle extinction after 1940. A circuitous 1991 argument by environmentalist Theodora Colborn reveals the importance of this claim in sustaining anti-DDT theology:[273]

> Although much of what is used to chronicle the fluctuations in bald eagle populations comes from gray literature and anecdotal sources, the above [text] reveals consistency in the reporting, indicating that by the early 1940s the bald eagle populations nationwide and in the Great Lakes area had recovered from turn of the century human exploitation and disturbance, but another population decline followed World War II.[274]

In the above quote, Colborn argues that recovery of bald eagles started with adoption of laws protecting eagles and ended with broad use of DDT. However, it is useful to examine the evidence behind her conclusion. The basis for her claim of the recovery and of consistency in the reporting is as follows:

> Broley, who had a summer home in Delta, Ont., recorded 15 bald eagle nests within a 20-mile radius of

[273] Theodora "Theo" Colborn is known by some as the Rachel Carson of the 1990s and was co-author of *Our Stolen Future*, a controversial book about chemicals, endocrine disruption, and human health.

[274] Colborn, T., "Epidemiology of Great Lakes bald eagles," *J Tox Envir Hlth*, vol. 33 (1991): 412.

his home from 1910 to 1950. In 1951 only one nest produced young. By 1957, he estimated that about 100 nests were left in southern Ontario (McKeating, 1985).

Weeks (1974) reported as many as 100 active and successful nests in southern Ontario around 1957 and probably twice that in 1900. Brownwell and Oldham (1985) noted that at the turn of the century reports were mixed. Reports cited one eagle's nest per ten miles, one per five miles, and even one per mile along the entire shoreline and tributary rivers to the Great Lakes. Brownell and Oldham (1985) searched Long Point Bird Observatory newsletters and found citings [*sic*] for five active bald eagle nests in the area in the early 1960s. In 1963, one active nest was reported; at the end of the decade no nests were reported."[275]

This is it, the text that forms the foundation of her argument. These observations suggest that the flimsiest of data will suffice as proof if one really wants to believe that eagle populations were recovering before the advent of DDT, but then declined with its use. In reality, these observations only document a continuous decline in eagle populations, not a recovery at all. Additionally, comprehensive data compiled in Bald Eagle Recovery Team reports for different regions of the United States do not support her claim that eagles were recovering before the advent of DDT.[276]

[275] Ibid.

[276] Murphy, T.M., et al., "Southeastern states bald eagle recovery plan," Apr. 19, 1989, http://ecos.fws.gov/speciesProfile/profile/speciesProfile.action?spcode=B008. (accessed Sept. 25, 2009);

Northern States Bald Eagle Recovery Team, "Northern states bald eagle recovery plan," http://ecos.fws.gov/docs/recovery_plan/060309b.pdf, (accessed Sept. 25, 2009); Chesapeake Bay Bald Eagle Recovery Team, "Chesapeake Bay bald eagle recovery plan". U.S. Fish and Wildlife Service. 1990, http://ecos.fws.gov/docs/recovery_plan/0604191.pdf, (accessed Sept. 25, 2009).

The claim of DDT harming eagle populations taps into a deeper, ongoing debate in wildlife-management circles about the relative importance of survival versus fecundity (the ability to reproduce) in wildlife management. Shooting, electrocuting, poisoning, and trapping of eagles, along with other causes of eagle deaths, relate to survival. Observations on cause of death for thirty-nine bald eagles in thirteen states from 1969 to 1970 reveal almost no role for DDT as a cause of bald eagle mortality. Of the thirty-nine eagles, only one had DDE (the primary degradation and storage product of DDT in humans) and PCB (polychlorinated biphenyls) levels that suggested that one or the other was the cause of death. Of birds autopsied, "Autopsy revealed that eighteen bald eagles were illegally shot; other causes of death were impact injuries, electrocution, emaciation, and infectious diseases."[277]

These causes of reduced survival are not limited to bald eagles. Of dead golden eagles collected during years of high DDT use (1964 to 1971), most deaths were due to shooting or contact with power lines. Other causes were trapping, collision with vehicles, and seemingly purposeful chemical poisoning (e.g., use of strychnine). Results from chemical residue analyses suggest that no eagles died from DDT exposure. Furthermore, many deaths were thought to be from natural causes (injuries from fights with other eagles, infectious diseases, etc.).[278] Poisoning of eagles occurred for the same reasons that at different points in time there have been bounties paid for dead eagles. Basically, eagles were seen as threats to farming and hunting productivity.

---

[277] Belisle, A.A., et al., "Residues of organochlorine pesticides, polychlorinated biphenyls, and mercury and autopsy data for bald eagles, 1969 and 1970," *Pesti Monit J*, vol. 6, no. 3 (1972): 133–38.

[278] Reidinger, Jr., R.F. and Crabtree, D.G., " Organochlorine residues in golden eagles, United States, March 1964 to July 1971," *Pesti Monit J*, vol. 8, no. 1 (1974): 37–43.

Reduced reproduction of bald eagles relates to fecundity. In arguing for the importance of fecundity, the anti-DDT movement pinned its hopes on establishing a relationship between two related claims. That DDT reduced eagle reproduction (productivity) is the first claim. That the decrease in eagle productivity caused eagle populations to decline is the second claim. The anti-DDT position seems to argue that if the first claim is true, then the second claim is also true, but others in the wildlife-management profession disagree.

In arguing for the claim that DDT endangered bald eagle populations, Colborn took stock of this underlying controversy:

> Longevity rather than fecundity is more important to maintain bald eagle populations (Young, 1968; Grier, 1974, 1980). Young (1968) hypothesized that sterility-causing insecticide exposure in short-lived, rapidly reproducing animals is an efficient approach to reduce insect populations. However, he also pointed out that "long-lived populations of slow breeders," such as bald eagles, are more strongly affected by direct mortality. Grier (1974) stated "that a decrease in reproduction may be far less important to an eagle population than an increase of comparable proportions in adult mortality." He later produced a model that "shows survival is far more critical to existence of a population" than reproductive success (Grier, 1980, p. 320).[279]

Colborn then engages in a discourse on speculative harms to bald eagle populations to support her claim that fecundity is more important to eagles than survival (or longevity). Yet, the theory and practice of wildlife-management principles have been verified in species after species and are widely accepted. Pablo Ferreras, a researcher who has studied the Iberian lynx, and his

279 Colborn, "Epidemiology of Great Lakes bald eagles," 409.

coauthors revealed as much in a recent study on the threat of extinction of this animal.

> The main factors driving to the extinction of endangered species populations are contraction and modification of their habitats and increased killing by humans (Caughley & Sinclair, 1994). This is especially true for large and medium-size mammalian carnivores in human-dominated landscapes, as they have important habitat and prey requirements and high risks of mortality of human origin.[280]

Ferreras' description of a large predator at top of the food chain, "in human-dominated landscapes," having "important habitat and prey requirements and high risks of mortality of human origin" certainly applies to the bald eagle. Ferreras highlights the contributions of mortality and habitat destruction to extinction, but he emphasizes the importance of mortality (that is, survival) even over habitat improvements:

> The relative importance of habitat degradation and increased mortality in the decline of specific populations must differ according to each situation. However, Woodroffe & Ginsberg (1998) recently postulated that high mortality in large carnivores inhabiting protected areas counteracts the positive effects of improving habitat, to the point that population size (related to the amount of preserved habitat) is a poor predictor of extinction. According to this hypothesis, conservation efforts that attempt to counter mainly stochastic processes (for instance, by managing habitat to enhance population size) are unlikely to avert extinction. Instead,

[280] Ferreras, P., et al., "Restore habitat or reduce mortality? Implications from a population viability analysis of the Iberian lynx," *Anim Conserv*, vol. 4, no. 3 (2001): 265–74.

> management of small populations of large carnivores should be focused on reducing mortality.[281]

The main forces that endanger wildlife are included in Grier's (quoted by Colburn) and Ferraras's analyses. Both conclude that reducing the mortality of wildlife is the most important intervention to prevent extinctions. Wildlife professionals use hunting seasons and bag limits (limits on numbers of animals that can be shot in one day or during a hunting season) to manage the population abundance of deer and other game animals. By applying these management principles, they prevent deer and other animals from becoming so abundant as to destroy their own habitats. There is broad agreement that the same principles, working in reverse order to prevent the shooting and taking of threatened species, are the most important intervention to prevent wildlife extinction. Such logic is apparent in the 1983 Northern States Bald Eagle Recovery Plan:

> Hypothetical modeling of bald eagle populations, using a variety of models and reproduction-survival schedules, shows that both reproduction and survival are important. Of the two, changes in survival have more impact on the population than similar changes in reproductive rates. Depending on survival, it is possible for populations with lower reproduction to do better than others with higher reproduction (Grier 1980). Reproduction of bald eagles appears to have increased following a period of reduced productivity from the late 1940's [*sic*] to the early 1970's [*sic*]. We do not know for sure what that means without concurrent survival information. The implications are 1) that variation in reproductive rates may not be quite as important as we formerly thought (although it is still important and cannot be ignored) and

[281] Ibid.

2) that we should pay more attention to survival and the habitat upon which survival depends."[282]

This conclusion questions the validity of the claim of DDT as a threat to bald eagles. Such conclusions are increasingly common in wildlife-management circles, as illustrated by a recent *National Geographic* article on bald eagles, which explains that "New research on the effects of DDT challenges the long-held belief that eggshell thinning was the primary cause of reproductive failure in birds. 'The thinning did occur,' said Buehler, "but it was probably not actually responsible for the reproductive failure."[283] (David Buehler is a wildlife professional who specializes in studies of bald eagles.)

Those who oppose all uses of insecticides argue vehemently against the conclusions of Buehler and other professionals. Preconceived and entrenched ideas that DDT was the culprit are seen in Colborn's writing about the role of productivity over survival as a cause of extinction. In fact, her synthesis of information about the importance of DDT and reduced productivity of eagles is highly biased and her arguments are weak, but she chooses these arguments because the role of productivity is critical to the cohesion of anti-insecticide propaganda. Regardless, the foundations of her arguments have been crumbling for many years.

At one time, the proposition that DDT and DDT alone caused thin eggshells, leading to egg breakage and reduced productivity, was so widely promoted by activists that it was beyond refute. Now, decades later, scientists are finally taking stock of the huge failings of the "evidence" and "science"

[282] Northern States Bald Eagle Recovery Team, "Northern states bald eagle recovery plan," 5.

[283] Winkler, R., "Bald eagle bounces back after decades of persecution.," *National Geographic News*, June 20, 2002, http://news.nationalgeographic.com/news/2002/06/0620_020620_bald eagle.html (accessed Apr. 13, 2009).

behind the DDT–bald eagle extinction fable. There is evidence that some relic populations of nesting bald eagles in the United States experienced reduced productivity and, in some localities, reduced productivity even persists today. Yet, there has been a fundamental failure to show that egg breakage was the actual cause of reduced nesting populations. This failure becomes extremely significant when we consider that the anti-DDT advocates have uniformly failed to show any other way in which DDT might have harmed eagle eggs and reproductive success. So, in the absence of solid data on the true impact of egg breakage on populations, the loud and boisterous claims of DDT causing the practical extinction of bald eagles begin to take on the flavor of scientific fraud. DDT opponents proclaim loudly and repeatedly that the cause-and-effect relationship between DDT, eggshell thinning, reduced eagle productivity, and risk of eagle extinction was proved beyond a shadow of doubt. Indeed, there has been great investment by DDT opponents in maintaining this charade, but this huge investment has not changed the truth of the cause-and-effect relationships. The charade has, however, cost much in terms of scientific enlightenment. To some degree it has distracted attention away from issues of how to manage bald eagle populations into the future.

As we will show, there are other, more meaningful explanations for how bald eagles were harmed, and whether DDT reduced the productivity of bald eagles is not as important a question as once thought. The important questions are (1) what were the real threats to bald eagles? and (2) how was the recovery of bald eagles in the lower forty-eight states actually accomplished? To answer these questions, we will examine how wildlife management officials worked quietly behind the scenes to save the bald eagle and, in the 1980s and 1990s, bring about a rapid recovery of bald eagles in the lower forty-eight states.

The Bald Eagle Protection Act of 1940

DDT opponents claim that bald eagles started to recover after (and even before) the Bald Eagle Protection Act of 1940. They further claim that wide use of DDT stopped the recovery and sent eagle populations into steep decline. As evidence for this, they present counts showing that numbers of juvenile bald eagles were low in the years after the Eagle Protection Act of 1940 was passed. They explain that the proportionately lower numbers of juveniles compared to adults were because DDT reduced the reproductive success of eagles. They further suggest that relative numbers of juvenile bald eagles increased after DDT was banned in 1972. In other words, they claim that banning DDT led to the recovery of bald eagles in the United States. As a unit, these claims make a neat and cohesive package that has been heavily marketed to reporters and politicians alike. The problem with this story is that it rests on highly selected data and a single interpretation that completely disintegrates on close inspection. This cohesive story had its beginnings in Rachel Carson's book.

Rachel Carson claimed that DDT was responsible for reducing eagle reproduction based on counts of juvenile bald eagles at Hawk Mountain Sanctuary:

> During the first years after the sanctuary was established, from 1935 to 1939, 40 per cent of the eagles observed were yearlings, easily identified by their uniformly dark plumage. But in recent years these immature birds have become a rarity. Between 1955 and 1959, they made up only 20 per cent of the total count, and in one year (1957) there was only one young eagle for every 32 adults.[284]

[284] Carson, R., *Silent Spring* (Aylesbury, Bucks, UK, Penguin Books: 1972), 114.

While Carson was correct in stating that it is easy to distinguish juvenile from adult bald eagles, it is not easy to distinguish a juvenile bald eagle from an adult golden eagle. Similarities between the two are old knowledge and were carefully described by Hornaday long before the 1940 act:

> It is unfortunate that this Eagle does not acquire its white head and tail until its fourth year. The head is fully feathered, and the name "Bald" refers solely to its white appearance. Up to three years of age, it is of the same general color as the golden eagle, and to distinguish the two species it is necessary to look at the lowest joint (tarsus) of the leg. If it is naked, the bird is a bald eagle; but if it is covered with feathers quite down to the toes, it is a golden eagle.[285]

With sufficient care, other differences can be used to separate the two, but the similarity of juvenile bald eagles to adult golden eagles allowed the hunting and taking of bald eagles to continue after 1940 because while the 1940 act outlawed the shooting and possessing of bald eagles, it did not protect golden eagles. And, because of its similarity to the golden eagle, the law also did not protect juvenile bald eagles. Thus, the crosshairs of hunting scopes passed over adult bald eagles to focus on the juveniles.

Carson was not the only person who noted that there were fewer juvenile bald eagles in the years before she published *Silent Spring*. Wildlife-management officials also apparently knew there was a decreasing survival of juveniles. In an ironic twist of timing, action was taken to amend the Eagle Protection Act to protect juvenile bald eagles in the same year Carson published *Silent Spring*. The Subcommittee on Fisheries and Wildlife Conservation of the House Committee on Merchant Marine and Fisheries drafted the amendment, and in 1962, the Eagle Protection Act of 1940 was

---

[285] Hornaday, W.T., *Hornaday's American natural history*, 16th ed (New York: Charles Scribner's Sons, 1935), 228.

amended to extend protection to the golden eagle. The amendment stated that "protection of the golden eagle will afford greater protection for the bald eagle, the national symbol of the United States of America, because the bald eagle is often killed by persons mistaking it for the golden eagle."[286] The reader should take special notice of word usage in the justification for that amendment. The authors did not state juvenile bald eagles "might be" or "could be" killed, they stated the bald eagle "is often killed by persons mistaking it for the golden eagle."

If hunting was the primary cause of reductions in bald eagle populations, as evidenced by the Bald Eagle Protection Act, then one could reasonably expect populations to increase as a consequence of little or no hunting. Indeed, there was a strong increase in numbers of bald eagles as a result of reduced hunting during WWII. In 1965, Walter Spofford, who worked in anatomy at the State University of New York Medical Center in Syracuse, reported on trends of bald eagles at Hawk Mountain.[287] He presented yearly count data from Hawk Mountain Sanctuary from 1935 to 1962. From 1935 to 1942, numbers of bald eagles varied from year to year but remained relatively constant at about fifty eagles per year. Counting was suspended during the war but was restarted in 1946. The count was just under fifty in 1946 and increased to about eighty-five in 1947. The counts increased in subsequent years and peaked in 1950 at about 140 eagles. The counts then began to decline during the next few years. From 1958 through the end of his data series, the counts declined slightly but were still around the pre-war levels recorded during the 1935 to 1942 interval.

---

[286] Indian Affairs: Laws and Treaties, Public Law 86-884, http://digital.library.okstate.edu/KAPPLER/vol6/html_files/v6p0989.html#mn1 (accessed Apr. 13, 2009).

[287] Spofford, W.R., "Hawk Mountain counts as population indices in Northeastern America," in *Peregrine falcon populations: Their biology and decline*, ed. J.J. Hickey (Madison: University of Wisconsin Press, 1969), 323–32.

The pattern of increasing numbers that evolved after the war reflects the time lag in the maturation of bald eagles. In other words, a bald eagle that hatches will not be differentiated from golden eagles until its fourth or fifth year. By and large, the increase in nesting success and, more importantly, improved survival of offspring and adults that occurred during the WWII hiatus in hunting would not be reflected in bald eagle counts until new birds had matured to their fourth year. Thus, those hatched in 1943 would be included in overall counts in 1947–48; those hatched in 1944, in 1948–49; and those hatched in 1945, in 1949–50. The 1950 count would be a more robust sum of increased survival and nesting success during the years of 1942–45. However, after 1950, the post-war impact of renewed and increased hunting pressure began to bring counts back down. Because of differential protection of adult bald eagles versus juvenile bald eagles, the post-war pressure affected juvenile birds more than adults.

People who campaign against DDT seldom mention actual bird-count data, such as that available from the Christmas Bird Counts or Hawk Mountain, because those data, both in general and in specifics, do not support their claims that DDT caused bird populations to crash with its broad use starting in the mid-1940s. Yet, in spite of the overwhelming evidence that shooting was the primary cause of declining populations of nesting eagles, reporters and science writers cling stubbornly to the belief that DDT was the major cause of wildlife harm, as revealed in a recent *National Geographic* article.

> The Bald Eagle Protection Act of 1940 prohibited shooting or otherwise harming the birds in the U.S. but didn't cover the pesticides that within a decade began to destroy eagles' eggs. By the 1960s only about 400 breeding pairs of bald eagles remained in the lower 48. "The trend . . . may well make it necessary for us to find a new national emblem," Rachel Carson warned in her 1962 masterwork, *Silent Spring*. The banning of DDT in

> 1972 and other measures launched an amazing comeback by the eagles, whose status changed from endangered to threatened in 1995.[288]

This version of the bald eagle story rests entirely on Carson's claim that reduced counts of juvenile eagles proved that DDT was reducing bald eagle productivity. The facts, however, refute this resilient story. In fact, this article was published shortly after the *National Geographic* article cited earlier that clearly stated the role that hunting and land-use changes played in bald eagles' demise.

Juvenile bald eagles started receiving their first meaningful legal protections from shooting in 1962. The relative numbers of eagles being shot, compiled in the eagle-recovery plan for southeastern states in April 1989, show an inverse pattern to that of increasing numbers of juveniles.

> Shooting has long been known as a major factor in the mortality of bald eagles. Reduction in shooting mortality is the current trend. The following figures indicate percentage of total reported mortalities caused by shooting mortalities: 62 percent from 1961–1965, 41 percent from 1966–1968, 46 percent from 1969–1970, 35 percent from 1971–1972, 25 percent from 1973–1974, 20 percent from 1975–1977, and 18 percent from 1975–1981. Thus, during a 21-year period, 25 percent of the documented bald eagle mortality has been caused by shooting. While this downward trend is encouraging, the current level of mortality is unacceptable and may be

[288]Eliot, John, "The bald eagle comes back from the brink," *National Geographic*, July 2002, http://ngm.nationalgeographic.com/ngm/0207/feature2/index.html?fs=animals.nationalgeographic.com (accessed Apr. 13, 2009).

> limiting, particularly in areas of remnant nesting populations. [literature citations removed][289]

These statistics should be considered a tip-of-the-iceberg phenomenon in that eagles can be shot and secretly ferreted away by hunters, and eagles taken in this way are not examined and are not included in statistical reports.

In contradiction to Carson's claims, actual counts of adult versus juvenile bald eagles at Hawk Mountain Sanctuary,[290] dating back to 1966, show increasing numbers of juveniles relative to adults from 1966 onward—years before DDT was banned. In the late 1980s juveniles became as numerous as adult eagles. Again, the increase in numbers of juveniles began in the 1960s, years before DDT was banned. These patterns show a strong relationship of increasing proportions of juveniles as numbers being shot declined. The logical conclusion is that juvenile bald eagles were commonly shot, or in some other way "taken," until the time the Eagle Protection Act was changed in 1962. Afterward, there was no loophole that allowed juvenile bald eagles to be shot: shooting them became strictly forbidden. As enforcement improved, the shooting of eagles gradually declined. As shooting declined, the age structure of national populations gradually changed to reflect the lowering of juveniles being shot more so than adults. Juvenile bald eagles eventually became more numerous than the adults.

Environmentalists charge that banning DDT led to the recovery of bald eagles in the United States.[291] This claim

---

[289] Bagley, F.M., et al., "Southeastern states bald eagle recovery plan," Apr. 19, 1989, p. 3.

[290] Hawk Mountain, "Conservation status report: bald eagle," http://hawkmountain.org/media/baldeagleCSR_June07.pdf (accessed Apr. 13, 2009).

[291] For instance, the U.S. Fish and Wildlife Service draws a direct link between DDT and eagle decline and infers that the banning of DDT led to the resurgence in eagle populations, stating that "today, Americans can see wild bald eagles in every state in the lower 48 and witness the

suggests that nesting populations continuously declined from DDT's widespread usage in 1946 up until the ban in 1972. As with so many claims of anti-insecticide activists, this claim is just not true. In fact, nesting pairs of bald eagles were increasing years before the ban. The following table of numbers of bald eagle pairs in the United States was compiled by the U.S. Fish and Wildlife Service (USFWS).

Figure 4-1: Number of Bald Eagle Pairs Counted in Lower 48 States, 1963–1994

Year	**Number**
1963	417
1974	791
1981	1,188
1984	1,757
1986	1,875
1988	2,475
1989	2,680
1990	3,020
1991	3,391
1992	3,747
1993	4,016
1994	4,452

U.S. Fish and Wildlife Service, Division of Endangered Species, Federal Register vol. 60, no. 133 (July 12, 1995): 36000–10; U.S. Fish and Wildlife Service, Final Rule to Reclassify the Bald Eagle From Endangered to Threatened in All of the Lower 48 States, 50 CFR Part 17. http://www.animallaw.info/administrative/adus60fr35999.htm

---

living proof of Rachel Carson's courage and determination." U.S. Fish and Wildlife Service, "Bald eagle numbers soaring," news release, May 14, 2007, http://www.fws.gov/news/NewsReleases/showNews.cfm?newsId=8B386C8A-F305-80F9-208693FCB6A533B4 (accessed Apr. 13, 2009). A WWF press release explicitly states, "Scientists attribute the bird's recovery to a 1972 ban in the United States on DDT, a pesticide that ruins the eggs of many birds, and strict protections under the Endangered Species Act and other environmental laws." WWF press release, "Statement on removal of bald eagle from endangered species list," June 28, 2007, http://www.worldwildlife.org/who/media/press/2007/WWFPresitem974.html (accessed Apr. 13, 2009).

These data show that numbers of bald eagle pairs increased from 417 in 1963 to 791 in 1974. DDT was de-listed for use in agriculture in January 1973 but it was used for an additional few years for exempted purposes and for public health. Additionally, DDT as a contaminant of Dicofol was still being used for agriculture into the 1980s. Regardless, the near doubling of mating pairs of eagles actually occurred during the critical years when DDT use was supposedly dragging bald eagles to the brink of extinction. Pairs of bald eagles increased even more rapidly after the ban, but this increase was not related to the de-listing of DDT for use in agriculture. It was related to two other factors. The first was the passage of the Endangered Species Act (ESA) of 1973, which funded re-establishing nesting in areas from which bald eagles had been eliminated. The other reason was the continual reductions in shooting mortalities of both juvenile and adult bald eagles.

If we step back in time to 1973, we find that the environmentalists' attacks on DDT and the claims about its impact on bald eagle populations scared the public and politicians. Their fear set the stage for political action, and action came in the form of banning DDT and passing the ESA. So, in the same year that use of DDT was stopped, the government passed legislation and opened public coffers for saving endangered species. Until the passage of the ESA, the primary source of funds for wildlife management came mostly from revenues derived from sports hunting.[292] The ESA suddenly made large sums of new money available for the USFWS and other agencies.

---

[292] "Hunters provide the primary support for all wildlife management, game, and non-game. License fees alone bring in approximately $500 million each year. In addition, excise taxes on hunting arms and ammunition provide nearly $200 million annually with sportsman support." Gun Muse Outdoors, "History of hunting," http://www.gunmuse.com/Main/Firearm+History/4-H+Program+Guide/Philosophy+Laws+and+More/History%20of%20Hunting (accessed Apr. 13, 2009).

At the end of a long stream of scary claims of DDT harm to wildlife, those campaigning against DDT struck gold with passage of the ESA. Programs for the recovery of the bald eagle and the peregrine falcon were two recipients of ESA funding, and high levels of funding continued for years thereafter. A 2005 report describes the level of ESA funding: "between 1989 and 2002, $5.196 billion was reported spent on individual endangered species. The USFWS and the National Oceanic and Atmospheric Administration (NOAA) spent $1.601 billion, while other federal agencies reported $2.909 billion and state wildlife agencies reported $0.694 billion in expenses." The report further stated that "bald eagle (*Haliaeetus leucocephalus*) conservation consumed $63.1 million (8 percent) of all USFWS spending."[293] In fact, in 1990 alone, the USFWS spent $3.5 million in ESA funds on the bald eagle and $2.9 million on the peregrine falcon.[294]

The Myth of DDT Endangering the Peregrine Falcon

Heavy funding for the recovery of peregrine falcons was also a result of fearful claims of DDT causing species extinctions. In general, the circumstances of reported declining populations of peregrine falcons were similar to the circumstance described for bald eagles. There were, however, some highly significant differences.

Peregrine falcons were reported to decline because of DDT during the years of heavy DDT use. Yet, there is almost no

---

[293] Male, T.D. and Bean, M.J., "Measuring and understanding progress in U.S. endangered species conservation," *Ecol Lett*, vol. 8 (2005): 986–92, http://www.environmentaldefense.org/documents/6168_MeasuringProgressinEndangeredSpeciesConservation.pdf (accessed Apr. 13, 2009).

[294] Abundant Wildlife Society of North America, "Endangered species act: flawed law," http://www.aws.vcn.com/flawed.html (accessed Apr. 13, 2009).

recognition that at the same time, there were no federal laws protecting the falcons. Falcon populations were very low but seemingly stable during the years when DDT protestors were claiming that DDT was causing falcon extinction. (See later for data on peregrine falcon numbers during the 1940s, '50s and '60s.) Claims about DDT as a cause of falcon endangerment are remarkably consistent in that they make almost no mention of the taking of peregrine falcons that occurred because falcons were in very high demand for the sport of falconry. A new book on the peregrine falcon by Jim Enderson,[295] an academician and falcon expert, describes in glorious detail the broad and intense pursuit and taking of peregrine falcons. Despite Enderson's gratuitous references to DDT as the cause of all harm to falcons, we recommend *Peregrine Falcon: Stories of the Blue Meanie* because it reveals much about the zeal of those who sought to possess a falcon. (That said, falconers did make key contributions to the rapid recovery of falcons in the United States.)

Furthermore, anti-DDT advocates ignored that the taking of peregrine falcons was an issue because the bird had a bad reputation—it kills pigeons and songbirds. In the 1935 edition of *Hornaday's American Natural History*, Hornaday reported that nine of twenty falcons examined had fed on songbirds. He described his solution to the falcon problem in graphic detail: "It can best be studied with a rope, a basket, and a chokebore shot-gun loaded with No. 6 shot. First shoot both male and female birds, then collect the nest, and the eggs or young, whichever may be present."[296]

This quote reveals how many people, especially songbird and pigeon enthusiasts, viewed the peregrine falcon. The pursuit of falcons for the sport of falconry, along with hunting and purposeful killing by pigeon fanciers, were the real causes of

[295] Enderson, J., *Peregrine falcon: stories of the blue meanie* (Austin: University of Texas Press, 2005).

[296] Hornaday, *American Natural History*, 228.

falcon endangerment. The falcon was not specifically protected from taking for any of these purposes and the protections that were in place were not enforced. As described in a 1983 Department of the Interior report, the only federal protection given to the peregrine falcon prior to 1970 was the Migratory Bird Treaty Act of 1918 and its amendments. However, as we saw with the bald eagle, there was no effective enforcement of this act. The peregrine falcon did not fall under specific protections until the 1970s when it was listed as endangered under the ESA. The falcon's lack of protection is amply illustrated in the many descriptions of falcon taking in Enderson's book.[297] It also describes how that all changed once the falcon was put on the endangered list.

The three subspecies of peregrine falcon all look the same. The Arctic peregrine falcon is more northern in distribution (Alaska, Canada), while Peale's falcon is more western in distribution (nesting from the Aleutian Islands east and south to Vancouver Island), and the American peregrine falcon occupies most of the lower forty-eight states. In the late 1960s, the latter subspecies was reported to have disappeared as a nesting species from Minnesota eastward and south to Georgia. It is perhaps no coincidence that its numbers decreased so much when humans populate this geographic region more densely than any other U.S. region. Peale's falcon, on the other hand, has never shown a population decline and was never listed as endangered. However, this subspecies migrates and would be expected to acquire contaminants just like other migrating falcons. The American and Arctic falcons were added to the lists of native (October 1970) and foreign endangered species (June 1970). In 1973, both subspecies gained strict protections from taking (shooting and trapping) thanks to the Endangered Species Act. In 1978, an estimated 2,900 falcons were produced in the

---

[297] Enderson, *Peregrine Falcon*, 273.

wild.[298] In 1983, the Department of the Interior proposed downgrading the classification of the Arctic falcon from endangered to threatened (the formal change in classification was announced April 6, 1984),[299] invoking the Similarity of Appearance provisions in changing levels of classification. This was done to simplify law enforcement because the birds were identical in appearance.

The details of these legal protections reveal the importance of taking as a source of falcon endangerment. World market value of peregrines ranged from $150 to $2,500 per bird from 1973 to 1983. This high value per bird was not new. The peregrine falcon was highly prized for trophies and falconry, and greatly despised by pigeon fanciers and songbird enthusiasts because falcons hunted these birds. As documented in Enderson's book and other literature, falcons were aggressively

---

[298] U.S. Fish and Wildlife Service (USFWS), "Proposed reclassification of the peregrine falcons in North America," Federal Register vol. 48, no. 41 (Mar. 1, 1983): 8796–8802; USFWS, "Endangered and threatened wildlife and plants; Reclassification of the Arctic peregrine falcon and clarification of its status in Washington and elsewhere in the conterminous United States," Federal Register, vol. 49, no. 55 (Mar. 20, 1984): 10520–26; USFWS, "Endangered and threatened wildlife and plants; Proposed rules to remove the peregrine falcon in North America from the list of endangered and threatened wildlife," Federal Register, vol. 63, no. 165 (Aug. 26, 1998): 45446–63; USFWS, "Endangered and threatened wildlife and plants; Final rule to remove the American peregrine falcon from the Federal List of Endangered and Threatened Wildlife, and to remove the Similarity of Appearance Provision for free-flying peregrines in the conterminous United States," Federal Register, vol. 64, no. 164 (Aug. 25, 1999): 46542–58; USFWS, "Post-delisting monitoring results for the American peregrine falcon (Falco peregrinus anatum), 2003" Federal Register, vol. 71, no. 198 (Oct. 13, 2006): 60563.

[299] Klinger, D., "Arctic peregrine falcon reclassified to 'threatened;' other protections for peregrines extended," Department of the Interior news release, Apr. 6, 1984, http://www.fws.gov/contaminants/pdf/historic/19840406.pdf

sought during all the years when DDT was being blamed for endangering falcons in the eastern United States.[300]

Many observations contradict the popular story that DDT was the driving force for endangerment of the American falcon. At one time, falcons in the eastern United States had the passenger pigeons as a vast and unending food source, but the passenger pigeon became extinct very early in the twentieth century. The problem, of course, is that we do not know the extent that the loss of its major prey set the stage for subsequent declines in numbers of falcons in the eastern United States. Regardless, the pigeon is the peregrine falcon's preferred prey, and loss of such a huge prey resource as that represented by the passenger pigeon was, without doubt, a major blow to peregrine falcon populations.

The 1983 USFWS report shows that once falcons were protected from taking, they began increasing throughout the 1970s and were showing strong increases in 1978. This was just five years after the birds became protected by the ESA. DDT was still present in the environment at that time, and was still being used as a major component of another broadly used insecticide, Dicofol. As described above, DDT was still being used as a result of EPA exemptions (mostly emergency exemptions) up through 1979. Falcon populations were doing well in Mexico even though Mexico was still making heavy agricultural use of DDT. Eight birds were sampled in Alaska in 1981, of which three had significant levels of DDT residues. The three birds produced eleven young and one cracked egg in their nest. This level of production was far above what is theoretically possible, at least according to exaggerated reports of anti-DDT advocates, for a bird with significant levels of DDT residues. Regardless, even with DDT still in use and still present in the broader environment, samples of migrating birds, as listed in the 1983 report, revealed that less than ten percent of birds had significant levels of DDT. Additionally, many migrating birds

[300] Enderson, *Peregrine Falcon*, 273.

were coming from areas south of the U.S. border where DDT was still heavily used in agriculture. This kind of evidence should bring into serious question the claim that DDT was the driving force that threatened the American falcon. What, then, actually endangered the peregrine falcon in the eastern United States?[301]

Joseph J. Hickey was an ornithologist in the Department of Wildlife Ecology at the University of Wisconsin in Madison. An activist who advanced the claim that DDT was responsible for the decline in the peregrine falcon populations, he organized an international peregrine falcon conference at the University of Wisconsin in 1965 and published a book on the conference proceedings in 1969.[302] Some conference contributors reached far beyond what their data would support and concluded that DDT and other insecticides were endangering the falcons. Others emphasized the inconsistencies in the data and cautioned against a rush to judgment about the threat of insecticides. However, Hickey and James Roelle, a wildlife ecologist, also at the University of Wisconsin in Madison, concluded at the end of the book that chlorinated insecticides "have led to persisting widespread reproductive failures and extirpation [elimination of peregrine falcons] over wide regions described in this volume."[303] So, in the end, Hickey got the result he wanted. All the statements cautioning against a rush to judgment and all the descriptions of other serious and continuing harms to falcons were put aside, leaving only DDT and other insecticides as the cause for the perceived crash of peregrine falcon populations.

---

301 U.S. Fish and Wildlife Service (USFWS), "Proposed reclassification of the peregrine falcons in North America," Federal Register vol. 48, no. 41 (Mar. 1, 1983): 8796–8802

302 Hickey, J.J., ed., *Peregrine falcon populations: Their biology and decline* (Madison: The University of Wisconsin Press, 1969).

303 Hickey, J.J., and Roelle, J.E. "Conference summary and conclusions," in Hickey, *Peregrine falcon populations.*

Hickey's 1965 conference signaled the beginning of the global campaign to stigmatize DDT as the singular cause for peregrine falcon problems in the United States. A review of papers presented at the conference reveals the "witch hunt" nature of the meeting and the separation between those with a biased, anti-DDT agenda and those who adhered to scientific evidence. Of the latter, three reports merit careful consideration.

One paper in Hickey's book reports on the extraordinary efforts needed to protect peregrine falcon nests. Joseph Hagar of the Massachusetts Division of Fisheries and Game reported on his assigned duties to protect fourteen aeries or eyries (falcon nest sites) after the Massachusetts legislature passed a statute that extended full protection to peregrine falcons in 1934.[304] Massachusetts was unusual in this regard because most states did not offer such protections. Hagar's observations were for the period 1935 to 1942. He explained,

> To guard these 14 eyries, I was substantially freed of other duties from mid-March to late June of each year, five district fish-and-game wardens were coached in the most effective ways of keeping part-time watch during the course of their other duties, and in the first 3 years, from two to four deputy wardens were hired to stand watch from early morning to late evening at certain sites where the risk of interference seemed high.[305]

Hagar stated that after the first year of such protections, "We had only one indication of a clutch [eggs in nest] taken by egg collectors." The extraordinary effort to protect the fourteen aeries was a realistic reflection of the high level of human disturbance inflicted on peregrine falcons.

---

[304] Hagar, J., "History of the Massachusetts peregrine falcon population", in Hickey, *Peregrine falcon populations*, 123–31.
[305] Ibid.

An examination of yearly productivity data for those highly protected aeries in Massachusetts reveals another important facet of peregrine-falcon reproduction. In spite of the claims that problems of falcon reproduction only began with broad use of DDT, Hagar's data reveal a relatively high reproductive failure rate for falcons in pre-DDT years. For example, even in the six best aeries, the percentage of pairs that failed to lay eggs varied from 0 to 66 percent. The failure rate was much worse for eight poorer nest sites, with 25 to 65 percent of pairs not laying eggs. Ninety five percent of nests with eggs produced one flying falcon. Overall, Hagar's data showed that even highly protected nest sites experienced low reproductive success.

Another paper worthy of consideration is "The extirpation of the Hudson River peregrine falcon population." This paper's senior author studied Hudson aeries from 1930 to 1960.[306] The paper reports continuous observations of a population of nests in a fifty-five-mile area over decades that included the years before DDT, the years covering WWII, and another fifteen years of DDT usage. The authors describe the gradual elimination of breeding falcons and report on the major factors that led to their elimination. Their critically important observations were as follows:

1. Hudson aeries were "heavily exploited by collectors, usually egg collectors."[307] The hobby of egg collecting did not decline in popularity until the mid-1940s.
2. "In the 1920s and 1930s the aeries began to be exploited by falconers."[308] "[A]ttrition by falconers did not take a heavy toll of the eyries until the mid-

[306] Herbert, R.A. and Herbert, K.G.S., "The extirpation of the Hudson River peregrine falcon population," in Hickey, *Peregrine falcon populations*, 133–54.

[307] Ibid., 135.

[308] Ibid.

thirties." Pigeon fanciers and gunners also took a heavy toll.

3. "World War II provided a temporary respite for the peregrines."[309] "Rationing of gasoline and the ban on pleasure driving made some of the aeries inaccessible, and the war effort removed many of the birds' human persecutors."[310] During the war years, the peregrine populations increased, but they started to decline again after the war.

The authors compiled a list of nest data for two intervals of years, 1931–40 and 1941–49.[311] As with Hagar's data, for occupied aeries (nesting sites with a pair of falcons present), the numbers of nests that produced young were always lower than the numbers of sites that failed. In other words, the reproductive failure rate in years before DDT was high. For example, from 1931–40, seventeen of thirty-two occupied aeries failed to produce young. The failure rate was higher from 1941–49, when twenty-seven of forty occupied aeries failed to produce young. Breeding pairs of falcons in the area dropped to zero in 1961.[312]

The authors commented on various possibilities for the falcons' extirpation. Regarding egg breakage as a possible consequence of DDT, they stated, "From 1930 to 1960 there was only one occasion on which broken bits of shell were seen on a nest ledge on the Hudson."[313] In their considerations of a possible role for insecticides as the cause for nesting failure, they stated, "We think that during the 1940s and 1950s other human pressures on these birds were such that extirpation was inevitable."[314] In summary, they wrote,

---

[309] Ibid.

[310] Ibid.

[311] Ibid., 136.

[312] Ibid., 147.

[313] Ibid.

[314] Ibid.

> Productivity in the 1940s was importantly affected by falconers who took 50% of the young as well as some incubating adults. Highway construction and blasting destroyed one nesting site in the 1950s and caused repeated desertion of eggs at other eyries. Unfavorable weather appeared to affect reproduction in 1 year, and shooting of the adults additionally contributed to the extirpation of this population.[315]

The last key report from the 1965 meeting was presented by Howard F. Young of the Department of Biology at Wisconsin State University in La Crosse. His paper, "The significance of the decline,"[316] compared the differential effects of increased mortality (lower survival) versus decreased reproduction. He concluded,

> It would appear to me that in a bird like the peregrine, mortality would be the most important of the effects compared here. And I think this is related to the fact that some of the population consists of the non-breeding subadults. If one turns to a bird like the bald eagle, which may not mature until the fifth or sixth year, and makes similar computations, one finds an even more striking difference between the effects. Mortality seems to take the eagle down in comparison at a faster rate; it is more important in the eagle than, say, sterility, for example. If one turns to a short-lived bird like the robin which breeds at 1 year of age and is a more rapid breeder, then you find that sterility will decrease the population faster than mortality.[317]

---

[315] Ibid.

[316] Young, H.F., "Hypotheses on peregrine population dynamics," in Hickey, *Peregrine falcon populations*, 513–19.

[317] Ibid., 518.

Young's assessments have been proven true over and over in wildlife literature. Indeed, his assessments can be viewed as basic principles of wildlife management. Survival is, for certain, the most important factor that could lead to the extinction of a bird like the peregrine falcon. Yet, this truth was silenced at the opening bell in the fight to characterize DDT as the single cause for falcon endangerment. Young's analysis shows that reproductive failures could result in a much greater impact on robins. Yet, we know heavy DDT use and intensive robin exposures to DDT did not even slow the growth of robin populations in the United States.

Participants in the 1965 conference offered a number of cautioning statements about what must have seemed to them to be a rush to judgment about DDT. W. H. Stickel, a wildlife researcher at the Patuxent Wildlife Research Center, said,

> Let us make sure that we are not missing the real problems—the real causes—by joining a single crusade. The recent decline [in falcons] is attributed to pesticides chiefly because it coincides in time with the advent of modern pesticides after World War II. This is awfully tricky in both logical and biological senses. Almost everything has changed or speeded up since 1945: industrial air and water pollution, development of nearly all waterfront property, suburbanization, motorboat travel in all waters, sale of telescopic sights on .22 rifles, contamination of roadsides with lead from gasoline, human travel into back country, even the disturbance that scarce hawks suffer from observers, falconers, and photographers . . . The long decline [in falcon populations] before 1945 suggests basic difficulties in habitat factors such as food supply, nesting sites, disturbance, disease, or pollution. Whatever the

> difficulties were then, they probably continue today, and very likely are more intense.[318]

It is such a shame that Hickey and his anti-DDT colleagues did not take seriously Stickel's reasoned analysis and warning. The conferees at Hickey's 1965 meeting ignored the fact that human actions such as hunting and habitat destruction were, and still are, the single greatest threat to wildlife. If the conferees had paid attention to that fact, as Stickel suggested they should, they would have realized that the threat had grown greatly from the 1940s through the 1960s. Human populations in the United States practically doubled between 1900 (76 million) and 1946 (140 million),[319] when DDT came into use, and grew by another 53 million from 1946 to time of the 1965 meeting in Wisconsin. That population increase alone was two thirds of the total U.S. population in 1900. If one accepts as fact that humans were the major threat to raptors (birds of prey, like the eagle and the falcon), then one can hardly ignore that the threat was growing rapidly during the 1940s, '50s, and '60s. Furthermore, the number of hunters grew greatly, as did numbers of falconers, after WWII. If we look back on how the falcon recovered once it received ESA protections, then we have to conclude that the human threat was neutralized only by virtue of outlawing the taking of falcons and placing severe penalties on anyone who dared break the law.

Enderson attended Hickey's 1965 conference in Wisconsin, and he wrote about the chemists invited to the meeting in his book. He noted, "it was ironic that the chemists were not of much help."[320] Enderson gave special demeaning attention to

---

318 Stickel, W.H., "General discussion," in Hickey, *Peregrine falcon populations*, 538.

319 U.S. Census Bureau, "Population: 1900 to 2002," *Statistical Abstract of the United States*, 2003, http://www.census.gov/statab/hist/HS-01.pdf. (accessed, Sept. 25, 2009).

320 Enderson, *The peregrine falcon*, 65.

one chemist, "Don Spencer, an employee of the research branch of the United States Department of Agriculture. Spencer would later sell his soul to the pesticide manufacturers and lobby against the eventual restriction of DDT."[321] Like the religious zealot who takes personal offense if anyone should speak against his religion, anti-DDT activists quickly resort to ad-hominem attacks on those who disagree with their position on insecticide use. That Spencer disagreed with Hickey and the other anti-DDT campaigners does not mean that he sold his soul to the pesticide industry.

Enderson described in detail the interests and activities of falconers. He described falconry as beginning in the United States in the late 1930s and early 1940s. A group called the North American Falconers Association was started in 1941, but he states that WWII then interrupted falconry for several years. Later, Enderson stated that by war's end in 1945, falcons had become broadly available: "It is safe to say that anyone who had established himself with the core of falconers in the eastern states could obtain a duck hawk [peregrine falcon]. Nestlings may have been harder to get because eyrie raiding was pretty intense."[322] He preceded his assessment about falcon availability in 1945 with the observation that

> in 1938, a phenomenal new source of peregrines was discovered. Bill Turner learned from his father that birds like those taken from local eyries could be found on the beach at Assateague Island, Virginia, in the fall. Turner's father produced a pair of peregrine feet to prove it. As it turned out, his hunting cronies found falcon shooting adequately challenging when the ducks weren't flying.[323]

---

[321] Ibid.

[322] Ibid., 135.

[323] Ibid., 134.

In his comments about the relationships between DDE and thin egg shells, he stated, "The unhatched eggs Craig and Berger collected in Colorado over the years tended to have thinner shells if the egg contents tested high in DDE. However, the relationship was not very tight."[324] He also reported,

> We found the usual case is the presence of both thick- and thin-shelled eggs in the same set (clutch). One eggshell from the clutch did not represent the rest. Further, shell fragments picked up on the ledge after the eggs hatched naturally showed the same bewildering variation in thickness. The vagaries of DDE and shell thickness confounded our attempts to link them with the number of young that successfully flew from the nest.[325]

In fact, this is a circuitous way of saying he found no association between shell thickness and production of young falcons. There is no evidence that DDT had any role whatsoever in the problems of peregrine falcons in the United States. Nonetheless, Enderson described how falcon-recovery plans were developed to meet USFWS requirements. He stated that the Alaskan plan was designed to "stamp out the use of DDT and similar pesticides in Mexico and countries farther south."[326] As it turns out, anti-DDT advocacy, through NAFTA, eventually succeeded in stopping all uses of DDT in Mexico, including use of DDT for control of malaria.

---

[324] Ibid., 216.

[325] Ibid.

[326] Ibid., 188.

DDT: No Threat to Bald Eagles or Peregrine Falcons

Since the 1983 report,[327] the American falcon has been reestablished in the eastern United States and in 2009, the peregrine falcon is no longer listed as an endangered or threatened species. As with the bald eagle, humans were responsible for bringing the falcon back to nest sites in the areas from which they had been eliminated, and removing DDT from the environment was not the cause for species recovery. Recovery of the American falcon was entirely brought about by protecting the birds from taking and by introducing the birds back into as many nesting sites as possible.

The bald eagle and peregrine falcon stories have been used to demonize DDT and to claim that banning DDT led to recovery of both species. The anti-DDT campaign was instrumental in opening public coffers to vast expenditures through the ESA. Yet, claims against DDT were not based on solid science. Furthermore, the stories behind those claims do not hold up to scientific or statistical scrutiny. This is revealed in a realistic evaluation of risks for bald eagles that was developed by the Illinois Natural History Survey (citations have been removed):

> The concentration of organochlorine pesticides, polychlorinated biphenyls, and mercury have been monitored in bald eagle carcasses and eggshells since 1966, [and while] nearly all carcasses contained residues of DDE and PCBs [most], although not all, died from other causes … [I]t, is not certain to what extent [eggshell thinning] is limiting reproductive success, [because the] most common cause of death is illegal shooting although this has decreased; [sperm development was not affected by being] fed a sublethal

[327] USFWS, "Proposed reclassification of the peregrine falcons in North America," Federal Register, vol. 48, no. 41 (Mar. 1, 1983)) :8796–8802.

> dose diet of DDT. Trichomoniasis has been recently reported in bald eagles that apparently fed on the parasite's host, rock and mourning doves, [however] treatment with emetryl tablets produced complete recovery; habitat loss to agricultural and industrial uses, logging, and recreational utilization of lakes and lakeshores coupled with human disturbance has resulted in a long term decline; loss of nesting sites is particularly critical; Sprunt found low productivity to be due principally to fewer pairs of eagles producing young at a lower rate.[328]

As we can see in this evaluation, bald eagles are subject to many risks. The same is true for the peregrine falcon. The threat from DDT, if any, is, without a doubt, the least of those risks. Yet, the anti-insecticide campaign has been so overwhelming that there has been only limited opportunity to focus attention on the true risks. Regardless, through the Endangered Species Act of 1973, programs for the recovery of nesting populations of bald eagles were funded and put into motion. With their success in focusing public attention on issues of insecticides, DDT opponents took the credit from the people who actually did save nesting populations of bald eagles and brought about a remarkable recovery of bald eagles in the lower forty-eight states. The results of those programs come from the hard and dedicated work of many people across decades of restoration efforts. The recovery was not based on cleansing DDT from the environment: it was based on solid wildlife-management methodologies.

Specifically, through research and innovation, new methods for fledging eagles and falcons in unoccupied nesting sites throughout the United States were developed and implemented. Fledging eaglets at many different sites has returned nesting

[328] Indiana Natural History Survey, "Bald Eagle, *haliaeetus leucocephalus*," http://www.inhs.illinois.edu/animals_plants/birds/ifwis/birds/bald-eagle.html (accessed Apr. 13, 2009).

eagles to many locations from which they had been eliminated. In the latest counts of nesting eagles, there are now more than 6,000 pairs in the lower forty-eight states. It is a marvelous accomplishment, and credit should go to those who did the hard work of implementation and recovery. Eagles have now been introduced back into many locations, and the work of recovery continues. None of these successes has anything at all to do with banning DDT: DDT did not threaten extinction of bald eagles in the first place, and the banning of DDT was not the cause for their recovery.

Summary

A campaign of fear is a powerful force; and through it, Carson and her followers grabbed public attention from the underlying issues that truly threatened bald eagles. Not once in the pages of *Silent Spring* are hunting and shooting, intentional poisoning, electrocution, trapping, collisions with vehicles, loss of habitats, pressure from human population growth, or infectious agents mentioned as a risk to eagle survival or cause of eagle mortality. Carson sidestepped the true threats to eagles in order to focus public attention on DDT. To elevate DDT to national prominence, she conjured fear through conjectures and speculations disguised as science.

Some people might discount the harm caused by misleading the pubic and politicians because they believe that the end result—the reduction of the spread of a persistent chemical in the environment—was worthwhile; in other words, the ends justified the means. This is a dangerous and naive argument. Persistent chemicals are not new to the natural world, and their persistence does not automatically make them harmful. In fact, the science shows that DDT's persistence is not harmful. But the real danger of the argument is that it condones the blatant and purposeful misuse of science in pursuit of ideology.

As with any campaign that is based on fear, when truth and objectivity are set aside, opportunities arise for horrible things to happen. As those who opposed any and all uses of DDT succeeded in closing DDT factories and changing public-health policies, they didn't save wildlife, but they did gradually and inexorably increase the risk of sickness and death for billions of poor people around the world. Through their campaign of fear, this group of environmentalists wrested control of critical public-health programs from those who labored to control the great diseases. A colossal public-health disaster has been the legacy of their efforts.

Chapter 5

The Making of a Public-Health Disaster

Although DDT had proven to be one of the most powerful weapons for the control of major diseases, the UN and intergovernmental agencies, such as WHO, turned against the insecticide that they once championed. A review of discussions in WHO executive board meetings shows clearly that this change of heart did not result from scientists' fears about the complex and dynamic process whereby insecticide-resistant genes are driven through mosquito populations, but from misrepresentations of several scientific issues.

The Beginnings of Environmentalism

In the mid-twentieth century, people's thralldom to disease started lifting, a change reflected in the allocation of Nobel Prizes during that time. Between 1939 and 1952, out of the eleven prizes awarded in medicine and physiology, five went to scientists who had made significant advances in controlling disease.[329] By 1960, the disease profile of the United States had

[329] No Nobel Prize was awarded in 1940, 1941, or 1942. The five awardees are: 1938—Gerhard Domagk for "the discovery of the antibacterial effects of prontosil" (a dye that was the progenitor of the sulphanomide drugs that saved so many lives before the advent of penicillin); 1945—Sir Alexander Fleming, Ernst Chain, and Sir Howard Florey for the discovery of penicillin; 1948—Paul Muller for the discovery of DDT; 1951—Max Theiler for his discoveries of how to combat yellow fever; and 1952—Selman Waksman for the

changed radically. Tuberculosis, typhoid, and other bacterial diseases could be treated with antibiotics. Typhus was conquered, and malaria was eradicated. Polio was in sharp decline as a result of the Salk vaccine, and in three years the introduction of mandatory vaccination programs would greatly curb standard childhood diseases such as measles, mumps, and rubella.

The decline of infectious disease had already begun in the late nineteenth century with the improvements in living conditions (e.g., better house construction, energy, and plumbing) and hygiene enabled by the industrial revolution. The industrial revolution afforded people more wealth, leisure, and time to consider their relationship to the natural environment. Thus, it is no coincidence that the environmental movement also began to form at this time. By mid-century, concerns such as Muir and Pinchot's about the effects of industry on the environment were becoming more and more common.[330] William Vogt argued in his book, *The Road to Survival*, that free enterprise was responsible for the despoiling of nature's resources, and that as the population outstripped resources, a "meeting at the ecological judgment seat" was at hand. Fairfield Osborne, a businessman and author of *Our Plundered Planet*, thought that the free-enterprise system could restrain resource-management abuses, but he was skeptical of technology as the answer to environmental problems, arguing instead for "the necessity of cooperating with nature."[331]

Even though there was a general optimism about the virtues of the recent advances in technology, there was also a strong element that was skeptical that these changes were not all they

---

"discovery of streptomycin, the first antibiotic effective against tuberculosis." See http://nobelprize.org.

[330] It would be foolish to deny the impact of political ideas on these movements. A strong socialist strain ran, and continues to run, throughout both the public-health and the environmental movements.

[331] Gottlieb, *Forcing the Spring*, 36–37.

were cracked up to be. Given the growing concerns about the effects of chemicals on human health and the liberal use of DDT in U.S. agriculture after World War II, it is perhaps inevitable that someone would link the "miracle" insecticide, DDT, to human illness.

The Greens Begin to Win; The Poor Lose

At this very moment, in dozens of countries around the world, malaria death snatches the newborn from the mother's breast. Fathers are helpless against a disease that takes their children, wives, and often, their economic wherewithal (e.g., through loss of employment, or loss of crops because of illness at critical times for planting or harvesting, and cost of medical care). In some instances, the pregnant woman and her unborn child are snatched away at the same time. Yet, the devastation of uncontrolled malaria cannot be measured only in terms of lives lost. The illness and death of family members translates into severe psychological and financial loss for those who remain. When a country's malaria program is reduced to just case treatment, as WHO recommended for many years, disease-control operations are reduced to dealing with only small components of the overall problem.

How is it possible that such harm and loss could occur, year after year in country after country, without an outcry or demand for political action? It occurs because the people most harmed are impoverished and politically voiceless. Their births and premature deaths are unnoticed by all except those in their immediate surroundings, who are also impoverished and voiceless. They are the have-nots of the world, and they live in countries that are also impoverished.

Citizens of rich countries cannot simply shrug off these problems as having nothing to do with them or being not their responsibility, because rich countries largely dictate what is done to control the great diseases in poor countries. The terms of assistance and public policies of rich countries, combined with the opposition of environmental groups, have led to truly harmful (or at best, ineffectual) foreign assistance to the world's poorest people. The malaria-endemic countries that chose to maintain disease-prevention programs and continue using DDT after its ban are those least dependent on foreign assistance.

Specific events propelled the world toward this outcome. The evolution of the environmental movement, with the growth of its financial resources and political influence and its encouragement of anti-insecticide activists, resulted in the banning of DDT in the United States and other developed countries—a ban that was politically motivated, that ignored scientific evidence, and that recommended the use of a more dangerous chemical, methyl parathion, in place of DDT. Enormous harm has been inflicted on human health as a consequence. So why did the public, science reporters, and even scientists readily accept so many of the claims against DDT?

Rachel Carson's Claims and Nature's Response to DDT

Public and scientific receptivity to claims against DDT stemmed from Rachel Carson's assertion that man's use of persistent and bioaccumulative chemicals, like DDT, was an assault on nature. What made her assertion so believable was her largely uncontested claim that the natural world had no defensive responses to such chemicals:

> The chemicals to which life is asked to make its adjustment are no longer merely the calcium and silica and copper and all the rest of the minerals washed out of the rocks and carried in rivers to the sea; they are the

> synthetic creations of man's inventive mind, brewed in his laboratories, and having no counterparts in nature.[332]

Carson mounted a specious argument that man-made chemicals were alien to the natural world, and environmentalists embraced her claims and trumpeted them loudly and successfully. Improved understanding, however, has shown that these chemicals are not alien at all. The fact that the natural world responds to DDT through degradation, sequestration (meaning that DDT is tightly bound to organic material), compartmentalization (meaning that it is mostly stored in fat cells of living organisms), and resistance mechanisms (described in chapter 1) shows that nature did, in fact, have preexisting and highly effective capacities for dealing with DDT-like compounds. To be fair, at the time that Carson was writing, knowledge of naturally occurring DDT-like chemicals was limited, but this does not excuse Carson's presentation of an unfounded belief as fact.

Carson also argued that when insecticides are used, insects become resistant, requiring stronger chemicals, and that this endless cycle of resistance shows that the chemicals shouldn't be used in the first place. Her argument is circular and defies science. In the face of resistance, we don't need stronger insecticides—we need ones with different modes of action. Bacteria have developed resistance to antibiotics and will continue to develop resistance, yet this is not used as an argument against these life-saving medicines. Rather, we argue for continued investment to develop new types of antibiotics.

But despite the flaws in Carson's arguments, they were embraced, enhanced, and repeated as unshakable truths through decades of environmental activism. Today, as revealed in World Health Assembly resolution 50.13 calling for reduced reliance on insecticides for disease control, Carson's followers not only seek to eliminate and stop production and use of DDT for

[332] Carson, *Silent Spring*, 7.

protecting people from malaria, they cast their net widely to encompass all insecticides, particularly compounds with atoms of chlorine, bromine, or fluorine.

DDT and the Infinite World of Natural Chemicals

DDT is an aromatic organic compound, which means that its parent structure is a hydrocarbon built mostly with rings of carbon atoms connected by alternating single and double bonds and bound to hydrogen atoms. Aromatics and their derivatives are ubiquitous throughout the natural world, and some (e.g., vitamin D) are absolutely essential for life.[333]

DDT contains five chlorine atoms, which replace five of the hydrogen atoms in the parent hydrocarbon. Chlorine is the eighteenth most common element on earth. Table salt contains chlorine—its scientific name is sodium chloride—and sea water is 2.7 percent by weight sodium chloride. This represents 78 million metric tons of sodium chloride per cubic kilometer of seawater. Chlorine, like iodine and fluorine, is a halogen. Organic molecules containing halogens are referred to as halogenated compounds. So we can broadly classify DDT as a halogenated aromatic hydrocarbon, or an organohalogen, or a chlorinated hydrocarbon.

---

[333] Purdue University, College of Science, "Naturally occurring hydrocarbons and their derivatives," http://chemed.chem.purdue.edu/genchem/topicreview/bp/1organic/hydro.html (accessed Apr. 14, 2009).

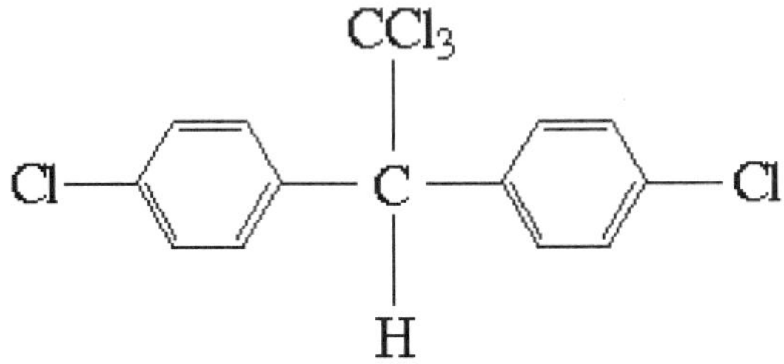

DDT

DDT is a synthesized product with 5 chlorine atoms.

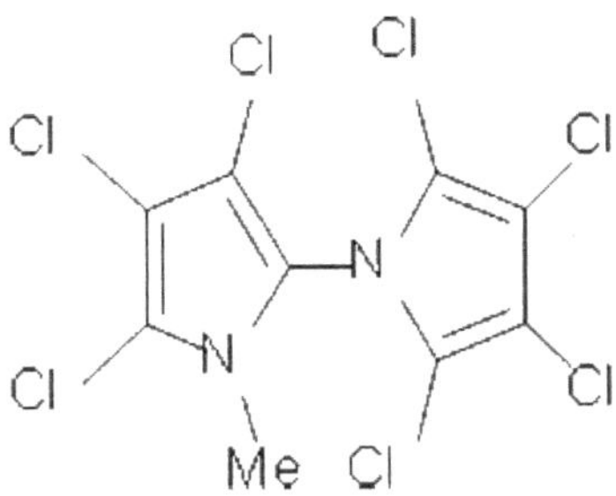

Q1

Q1 is a natural product with 7 chlorine atoms.

The fact is, many living organisms produce molecules like DDT for the same reasons we do: for defense and other purposes. DDT is a man-made chemical, but it is simple and easily produced. In contrast, and ironically for many activists that campaign against the use of chlorine and chlorinated compounds, many natural chemicals are far more complex and even more highly chlorinated than DDT.

As stated earlier, while Carson was writing and researching, our knowledge of the natural organohalogens was limited. In 1994, Dr. Gordon Gribble, an internationally recognized chemist, explained this in quoting a statement made by scientist Dr. L. Fowden in 1968:

> "Present information suggests that organic compounds containing covalently bound halogens are found only infrequently in living organisms."[334]

Gribble went on to say that "although this observation was undeniable 25 years ago, it is nothing but a myth today."[335] But, as Gribble observed in a 1992 paper, that myth persisted long after it was disproven:

> The Science Advisory Board to the International Joint Commission on the Great Lakes stated recently: "There is something nonbiological about halogenated organics (excluding iodinated compounds) . . . Chemicals [that] do not occur naturally . . . are often persistent, since there are often no natural biological processes to metabolize or deactivate them" (1). Not only are these statements incorrect, but, coming from a "Science Advisory Board," they reflect a disturbing ignorance of the chemical composition of our environment and of [the] biochemical processes therein. Similarly, in a 1990 *Science* article, a well-known cancer researcher was quoted as saying: " . . . some types of synthetic compounds, including halogenated hydrocarbons such as PCB, are not found in nature" (2). To the contrary, the sheer number of known naturally occurring halogen-containing chemicals is . . . vast.[336]

Dr. Walter Vetter, an internationally recognized chemist at the University of Hohenheim, Germany, is arguably one of the leading experts in the discovery of natural organohalogen compounds. We now know that such chemicals are natural

---

[334] Gribble, G.W., "The natural production of chlorinated compounds," *Environ Sci Tech*, vol. 28 (1994): 310A–319A.

[335] Ibid.

[336] Gribble, G.W., "Naturally occurring organohalgen compounds—a survey," *J Nat Prod*, vol. 55, no. 10 (1992): 1353–93.

products, that living organisms make use of these compounds in myriad ways,[337] and that they can be present in greater abundance than even man's synthetic compounds.[338]

Vetter was the first to characterize some of the major persistent organohalogens in the blubber of ocean-going mammals. He also analyzed museum specimens of euglossine bees (a type of orchid bee found in Brazil) that had harvested DDT. The environmentalist might interpret the harvesting of DDT as an error in the bees' selection of compounds—an error that would ultimately result in the bees' death. In fact, the bees are naturally equipped with specific abilities to detect DDT from long distances, fly and land on DDT surfaces, then collect and store relatively large amounts in internal pouches of their hind legs. Unlike mosquitoes and other insects that are repelled or killed, the bees are attracted to it and not harmed by their DDT exposures.[339] In Vetter's chemical analyses of the bees, he discovered three unknown natural products. These three are the most highly chlorinated natural compounds yet discovered, and his findings suggest that the male bees harvested DDT as a building block for the synthesis of even more highly chlorinated compounds. While the function of the unknown natural compounds remains open to discovery, the chemistry of the euglossine bees illustrates how living organisms are fully

---

337 Vetter, W., et al., "Anthropogenic and natural organohalogen compounds in blubber of dolphins and dugongs (Dugong dugon) from northeastern Australia," *Arch Environ Contam Toxicol*, vol. 41, no. 2 (2001): 221–31; Vetter, W., et al., "Sponge halogenated natural products found at parts-per-million levels in marine mammals," *Environ Toxicol Chem*, vol. 21, no. 10 (2002): 2014–19; Moore, M.R., et al., "Trace organic compounds in the marine environment," *Mar Pollut Bull*, vol. 45 (2002): 62–68.

338 Gribble, G.W., "Naturally occurring organohalogen compounds--a comprehensive survey," *Fortschr Chem Org Naturst*, vol. 68 (1996): 1–423.

339 Roberts, D.R., et al., "Male *Eufriesia purpurata*: a DDT-collecting euglossine bee in Brazil," *Nature*, vol. 297, no. 5861 (1982): 62–63.

integrated into a complex world of persistent DDT-like chemicals.[340]

The story of these natural compounds is still unfolding. But studies have already shown that there is a rich world of natural chemicals, some of which persist and pass through the food chain just like DDT, PCBs, and dioxins.[341] As Vetter stated in an abstract of a recent paper,

> A wide range and steadily increasing number of halogenated natural products (HNPs) is detected in marine organisms that are not the natural source of these compounds but which have accumulated these HNPs in a similar way as is known to occur with anthropogenic halogenated pollutants such as PCBs and DDT. The HNPs have aromatic, aliphatic, and heterocyclic spines and are bromiated, chlorinated, or mixed halogenated (CL and Br). The exact isomer structures of HNPs are often closely related to the anthropogenic POPs [DDT, etc.], and for some compounds both natural and anthropogenic sources are likely to exist. Some of the HNPs are nonpolar, persistent, and can thus be found even in marine mammals and birds of prey . . . Each of these compounds has been detected in higher trophic biota with concentrations exceeding 1mg/kg.[342]

What Vetter describes in this abstract is entirely contradictory to what the public has been led to believe by those who continue Carson's campaign against DDT. In his reference to higher trophic biota, he is describing large animals that feed on smaller

---

[340] Vetter, W. and Roberts, D.R., "Revisiting the organohalogens associated with 1979-samples of Brazilian bees (Eufriesea purpurata)," *Sci Tot Environ*, vol. 377, nos. 2–3 (2007): 371–77.

[341] See appendix 3 for further information about chemical threats to wildlife.

[342] Vetter, W., "Marine halogenated natural products of environmental relevance," *Rev Environ Contam Tox*, vol. 188 (2006): 1–57.

prey, such as eagles that feed on waterfowl or fish. Trophic level relates to the classic pyramid of numbers. So killer whales feed on seals, seals feed on large fish, large fish feed on small fish, and so on. In terms of numbers, then, there are fewer killer whales than seals, fewer seals than fish, fewer large fish than small fish, and so on.

DDT-like natural chemicals are produced and used by living organisms and have accompanied the biological world in every step through evolutionary time.[343] In another study, Vetter looked for the presence of persistent, lipophilic (fat soluble), and bioaccumulative (stored in fat cells) compounds in five species of Arctic sponges. Vetter and his coauthor found that the sponges were relatively uncontaminated.[344] They found no PCBs and only traces of DDE in the sponges. Each sponge contained at least thirty-five halogenated natural compounds and at least fourteen were common in all sponges. These natural compounds were highly abundant, with concentrations varying from nanograms/kilogram to milligrams/kilogram (air-dried weights). Some of these chemicals were subsequently found in seal blubber from both the Arctic and the Antarctic, illustrating movement of these natural products through the food chain.

Even before this study, Vetter had reported other natural organohalogens accumulating through the ocean food web by passing from smaller to larger animals (from lower trophic levels to higher trophic levels), noting that "biota [living organisms] can produce persistent organic chemicals that accumulate to substantial concentrations in higher trophic organisms."[345] In another study, he presented proof that such chemicals are produced in nature and that they do indeed

[343] Gribble, G.W., "The diversity of naturally produced organohalogens," *Chemosphere*, vol. 52, no. 2 (2003): 289–97.

[344] Vetter, W. and Janussen, D., "Halogenated natural products in five species of Antarctic sponges: compounds with POP-like properties?," *Environ Sci Tech*, vol. 39, no. 11 (2005): 3889–95.

[345] Vetter, "Marine halogenated natural products," 6.

accumulate to substantial concentrations in higher trophic level organisms.[346]

It is now known that the halogenated compounds are found in both terrestrial and aquatic organisms, but the greatest production of these compounds seems to occur in the oceans. Gribble explains,

> The oceans are the single largest source of biogenic organohalogens [compounds containing chlorine, fluorine, bromine, etc. that are produced by living organisms], which are biosynthesized by myriad seaweeds, sponges, corals, tunicates, bacteria, and other marine life. Terrestrial plants, fungi, lichen, bacteria, insects, some higher animals, and even humans also account for a diverse collection of organohalogens.[347]

Organohalogens are, however, found everywhere and are not limited to ocean environments. Anders Grimvall, a professor of statistics and environmental sciences at Linköping University in Sweden and the author of over one hundred scientific studies, states that "organic forms of halogens are ubiquitous in rain and snow."[348] In discussing the organochlorines in precipitation, Grimvall states that anthropogenic (man-made) chemicals in precipitation are about "three orders of magnitude less" than the absorbable organic halogens in rainfall. Regarding organochlorines, he states that they are abundant in surface waters and the majority are of natural origins.[349]

---

[346] Vetter, "Marine halogenated natural products," 7.

[347] Gribble, "Naturally occurring organohalogen compounds."

[348] Grimvall, A., "Natural organochlorines in precipitation and surface waters. In The natural chemistry of chlorine in the environment. A series of views by a panel of independent scientists." World Chlorine Council, 1999, 7–8.

[349] Ibid.

The public, scientists, and policy makers should pay close attention to the unfolding discoveries of scientists like Vetter, his colleagues, and others who are researching the natural organohalogens. Their work is leading to new and constructive interpretations of the chemistry of life, and opening the door to an entirely new realm of natural chemicals that can further serve mankind, either as models of natural cycles or, potentially, as new chemical tools against infectious and/or chronic diseases. For example, some of these natural chemicals behave like DDT: they are persistent and are present in the food chain. These chemicals have been with us through evolutionary time, and living organisms have developed natural responses and capacities to deal with such chemicals. As another example, the immensely valuable antibiotic vancomycin is a complex chlorophenol—a molecule that contains chlorine and several aromatic rings[350]—but it is much more than this. Vancomycin is a natural product and an immensely powerful weapon in the fight against human infections. Likewise, DDT is much more than a chlorinated hydrocarbon with aromatic rings. It is, in fact, still the most powerful substance known for protecting humans inside their homes against malaria.

DDT and Human Health

Over the decades of DDT use, a wide variety of claims have been made about DDT harm to human and animal health. Some claims have been truly absurd. An endless cycle of these claims is nonetheless fed and sustained by the public's appetite and its readiness to accept them without scientific evidence. For example, a 1940s claim attributed to DDT, without a shred of evidence, a mysterious cattle disease known as X-disease (discussed further in appendix 4). In this anti-science context, the pressures against DDT continued to build, culminating in the

[350] Gribble, G.W., "Natural organohalogens," *J Chem Edu*, vol. 71, no. 11 (Nov. 1994): 907–11.

1972 EPA hearings that banned the chemical. Many qualified scientists stepped forward in efforts to expose the misrepresentations (or in some cases, complete lack) of science that eventually led to those hearings. One such scientist was an internationally recognized expert on DDT toxicity, Dr. Wayland Hayes. He observed that

> the safety record of DDT for man is truly remarkable—several hundred thousand metric tons per year have been used for agriculture, forestry, public health and other purposes. People have been exposed in almost every way conceivable. For typhus control, the entire populations of villages and large cities have had the 10% powder blown into their clothing as they wore it. For malaria control, millions of men, women, and children have had the interior walls of their homes sprayed year after year at the rate of 2 g/m^2. For control of the yellow fever mosquito, DDT has been added directly to community water supplies at a concentration of 1 ppm and to individual water containers in the home at concentrations up to 5 ppm . . . For food protection, such a variety of plants and animals eaten by people have been treated that, apparently without exception, every person sampled in recent years in Europe, Asia, Africa, Australia, and the Americas contains a trace of DDT in his blood and in every tissue. Some of those who manufacture, formulate, or apply DDT have had no other work for over 20 years and all of them have greater exposure than the general population of their countries. Yet, in spite of the prolonged exposure of the entire population of the world and the heavy occupational exposure of a substantial number, the only

confirmed cases of poisoning by DDT have been the result of massive accidental or suicidal ingestion.[351]

One might argue that despite this safety record, DDT is still bad because it is classified as a carcinogen. Classification of carcinogen activity comes from tests on rodents given high doses of DDT. There is controversy, however, regarding the utility of such data for defining human cancer risks.[352] Furthermore, the animal data have allowed the anti-DDT literature to focus on DDT as a man-made carcinogen while ignoring a great variety of natural chemicals that show even greater carcinogenic potential.

The relevance of natural carcinogens to the risk of human cancers has been elevated to serious debate, and formal evaluations have revealed some startling findings. A high-level study in 1996 by the National Academy of Sciences found that "the human cancer risk from naturally occurring substances in the diet exceeds that of synthetic dietary carcinogens."[353] This finding is consistent with the views of some notable experts. Dr. Bruce Ames is one such expert. He is emeritus professor of biochemistry and molecular biology and director of the National Institute of Environmental Health Sciences Center at the University of California, Berkeley. He is a member of the National Academy of Sciences and has served on their

---

351 Hayes, "The degree of hazard to man of DDT as used against malaria," 1.

352 Wildavsky, A., *But is it true?: A Citizen's Guide to Environmental Health and Safety Issues* (Harvard University Press, 1997).

353 Committee on Comparative Toxicity of Naturally Occurring Carcinogens, Board on Environmental Studies and Toxicology, Commission on Life Sciences, National Research Council. "Carcinogens and anticarcinogens in the human diet: A comparison of naturally occurring and synthetic substances" (Washington, DC: National Academy Press, 1996), http://www.nap.edu/openbook.php?chapselect=yo&page=309&record_id=5150&Jump+to+Specified+Page.x=11&Jump+to+Specified+Page.y=16 (accessed Apr. 14, 2009).

Commission on Life Sciences. He has also served as a member of the National Cancer Advisory Board of the National Cancer Institute. He collaborated with a colleague, Dr. Lois Swirsky Gold, on a 1998 report on the misconceptions about environmental pollution, pesticides, and the causes of cancer. In the report, Ames and Gold surmise that there is no cancer epidemic, other than lung cancer as a result of smoking, and that allowing for age cancer mortalities have actually declined by sixteen percent since 1950. They opine that the major causes of cancer are smoking, dietary imbalances, chronic infections, and hormonal factors that are primarily influenced by lifestyle. These experts emphasize that

> regulatory policy that focuses on traces of synthetic chemicals is based on misconceptions about animal cancer tests. Recent research indicates that:
>
> a) Rodent carcinogens are not rare. Half of all chemicals tested in standard high dose animal cancer tests, whether occurring naturally or produced synthetically, are "carcinogens";
> b) There are high-dose effects in rodent cancer tests that are not relevant to low-dose human exposures and that contribute to the high proportion of chemicals that test positive;
> c) The focus of regulatory policy is on synthetic chemicals, although 99.9 percent of the chemicals humans ingest are natural. More than 1,000 chemicals have been described in coffee: 28 have been tested and 19 are rodent carcinogens. Plants in the human diet contain thousands of natural pesticides that protect them from insects and other predators: 63 have been tested and 35 are rodent carcinogens.[354]

[354] Ames, B.N. and Gold, L.S., "Misconceptions about environmental pollution, pesticides and the causes of cancer," NCPA Policy Report

Ames and Gold also emphasize that "there is no convincing evidence that synthetic chemical pollutants are important for human cancer."

Ames and Gold state that regulatory policy is driven by certain misconceptions about relationships between environmental pollution and human disease, particularly cancer. They characterize the following four statements as misconceptions:

- Environmental synthetic chemicals are an important cause of human cancer.
- Cancer risks to humans can be assessed by standard high-dose animal cancer tests.
- Synthetic chemicals pose greater carcinogenic hazards than natural chemicals.
- The toxicology of synthetic chemicals is different from that of natural chemicals.[355]

The take-home message here is that modern regulatory policies are based on seriously flawed concepts. Just as WHO ignores the scientific basis of how DDT acts to control malaria, the vast mechanisms of the regulatory process continue to ignore the weight of scientific evidence about high-dose animal tests and the risk of synthetic versus natural chemicals. The requirements for better perspective and more context in the regulatory process of potential carcinogens are also important in evaluating estrogenic activities of synthetic and natural chemicals.

Anti-DDT advocacy continuously highlights the role of DDT as a xenoestrogen (an environmental chemical that mimics estrogen). As an estrogen mimic, the anti-DDT propaganda holds that it has adverse affects on human reproduction.

No. 214, Mar. 1998, http://www.ncpa.org/pub/st214 (accessed Apr. 14, 2009).

[355] Ibid.

However, DDT is a weak environmental xenoestrogen. Many pervasive natural products are stronger xenoestrogens. John Katzenellenbogen in the department of chemistry at the University of Illinois describes the "structural pervasiveness of estrogen activity" as follows:

> A remarkably wide variety of chemical structures, produced both by nature and by man, have estrogenic activity. While the molecular details of this structural pervasiveness of estrogenic activity are not known, it is less surprising if one considers that assays for estrogenic activity are extremely sensitive and can operate over a potency range of 10 million, that most estrogens share the common structural motif of a phenol or the functional equivalent of a phenol, and that most estrogens are lipophilic and thus can be resistant to metabolism and undergo bioconcentration.[356]

In other words, we can detect chemical activities that are not important at the level of the living organism, in particular for an organism that is as large and complex as a human being. Katzenellenbogen goes on to emphasize that beneficial as well as undesirable effects can result from estrogen-active compounds. He points out that many natural products from plants and microbes are active estrogens. As stated in a 1998 EPA panel report of the Risk Assessment Forum on environmental endocrine disruption, "dietary exposure to xenoestrogens derived from industrial chemicals is minimal compared with estrogen equivalents from naturally occurring bioflavonoids." The panel concludes that "exposure to a single xenoestrogenic chemical at current environmental

[356] Katzenellenbogen, J.A., "The structural pervasiveness of estrogenic activity," *Environ Hlth Pers*, vol. 103, supplement 7 (1995): 99–101.

concentrations is probably insufficient to evoke an adverse effect in adults."[357]

These recent findings show ever-more clearly that living organisms are immersed in halogenated and persistent chemicals that are produced by other living organisms. These chemicals are abundant within both marine and terrestrial food webs. There is no meaningful difference between the biological activities of man-made chemicals, like DDT, and those of many natural compounds. There should be no doubt that life has evolved in close and interactive association with such chemicals. In fact, there should be no doubt that life itself is dependent on such chemistry.

The Environmental Protection Agency and DDT: Politics Trumps Science

As announced in an article in *Science* magazine in 1969,[358] the political agenda of the Nixon administration was to ban most uses of DDT and eventually extend the ban to other pesticides considered dangerous. At that time, the U.S. Department of Agriculture (USDA) was responsible for regulatory control of DDT. As described in the *Science* article, the USDA did not subscribe to the philosophy that there were adequate substitutes for all agricultural uses of DDT. However, President Nixon had promised to eliminate all but the essential uses of DDT by the end of 1970.[359]

---

[357] Crisp, T.M., et al., "Environmental endocrine disruption: An effects assessment and analysis," *Environ Hlth Pers*, vol. 106, supplement 1 (1998): 38.

[358] Kramer, J.R., "Pesticide research: Industry, USDA pursue different paths," *Science*, vol. 166 (Dec 12, 1969): 1383–86.

[359] Flippen, J.B., *Nixon and the Environment* (Albuquerque: University of New Mexico Press, 2000), 44.

It must have become obvious that little progress would be made until regulatory control was removed from the USDA and passed to another agency that would accommodate Nixon's political agenda. Thus, in 1970, the Nixon administration created the Environmental Protection Agency (EPA) and transferred regulatory control of pesticides from the USDA to the EPA. Not surprisingly, regulatory control of DDT was the first task of Nixon's newly created agency. In 1972, the EPA banned most uses of DDT in the United States. This was a political act to both gain approval from the burgeoning environmental community and create a political "victory" for the new agency. The ban, however, had no scientific basis, and the EPA replaced DDT with a more highly toxic and controversial chemical, parathion.

In the preface and in chapter 1, we touched on the issue of environmental groups opposing DDT because it was extraordinarily effective in saving lives and allowing increased fertility. It is a basic fact that any method of disease control that prevents the deaths of children and reproductive-age adults will aid population growth. For this reason, an anti-human, anti-population-growth ideology within the environmental movement believed that preventive measures to control malaria and other diseases should be avoided because people would be more likely to survive and reproduce. The degree of crossover between the two ideologies—the opposition to growth of human populations and the opposition to the public-health use of DDT—was extensive. Though many people who consider themselves environmentalists would never wish harm on others, the anti-population-growth movement campaigned tirelessly to rid DDT from public-health programs.

To put these issues into perspective, once more we need to look back at a sequence of events and major players in the environmental movement. WHO's malaria-eradication program (described in chapter 2) was portrayed as global in scope. Yet few people seem to know that most of Africa was never really part of that "global" program. Shockingly few countries in Africa ever had a true national program of malaria control, let alone a program of malaria eradication. The reported reason for not creating eradication programs in most African countries was the lack of health infrastructure needed for implementing the countrywide spraying of houses. In lieu of creating countrywide programs, pilot studies were conducted to test whether eradication was possible in selected countries. A pilot study was conducted in Uganda in the late 1950s. Spraying of houses proved remarkably successful and the investigators concluded it should be possible to eradicate malaria in Uganda.[360] However, the pilot study did not result in a countrywide eradication program. In fact, it resulted in no program at all. In contrast, Nepal had no national malaria program and no infrastructure for a countrywide program, yet USAID created an eradication program there. The program in Nepal started around the same time the pilot study was initiated in Uganda. No pilot study was performed in Nepal—USAID simply went in, created infrastructure, and implemented a countrywide program. That most countries of Africa were not provided that kind of assistance is troubling.

In 1965, an important and timely book by Peter Newman, *Malaria Eradication and Population Growth*, was published.[361] This book examined the relationships of malaria eradication and population growth in Ceylon (Sri Lanka) and British Guiana (Guyana). The purpose of the analyses was to determine if WHO's malaria-eradication program was responsible for the explosive growth of human populations that accompanied the success of malaria eradication programs. In answering this question, Newman surmised that malaria eradication contributed "about 60% of the postwar acceleration of population growth in Ceylon, and about 40% in British Guiana, so that this single

[360] Zulueta et al., "Malaria eradication pilot project."

[361] Newman, P., *Malaria eradication and population growth. With special reference to Ceylon and British Guiana* (Ann Arbor, MI: Braun-Brumfield, Inc., 1965).259pp.

factor did play a very major role."[362] Newman's work provided proof of what people already believed: use of DDT was saving lives and allowing, at least for the short-term, increased growth of human populations. However, looking back on subsequent events, we find that population growth in countries that eradicated malaria and then underwent significant economic improvements later experienced a slowing of population growth. This happened in the United States and almost every other advanced economy that experienced rapid population growth with the control of infectious diseases in the first half of the twentieth century, but then saw a decline in human fertility rates as wealth, education, and standards of living increased.

Newman admits that a question often asked in the 1960s was, "Is malaria eradication a good thing, bearing in mind its contribution to the population explosion?" The wording of his question, "population explosion," was indicative of an already-popular belief that human populations were growing at unacceptable rates. His own answer to that question is insightful:

> I would reply that this question cannot be answered scientifically, since it depends on one's judgment regarding the value of human life. My own feeling is that no one should take it on himself to withhold such effective, and marvelously cheap, methods of improving the general health of a people and, indeed, that the campaigns should be pressed with considerable vigor, since we now know that the indirect effects can be very great.[363]

His answer to the question would not be the final word on the use of a chemical that allowed growth of populations while protecting public health. As described in a *Dartmouth Science History* article, the EDF's Charles Wurster moved from

[362] Ibid., 3.

[363] Ibid., 6.

Dartmouth to the State University of New York at Stony Brook.[364] He joined the Brookhaven Town Natural Resources Committee, which worked to persuade Suffolk County (Long Island) officials to stop using DDT. The committee filed suit and the use of DDT was outlawed.[365] In the hope of widening the ban on DDT, the ten-member committee in October 1967 incorporated the Environmental Defense Fund (EDF).[366] The EDF decided to focus its efforts on litigation and gained media attention through court action against DDT in Michigan and Wisconsin.[367]

Population Growth and DDT

At some point in the 1960s, David Brower, who was the executive director of the Sierra Club,[368] and who, in an interview with the *San Francisco Chronicle* in 1998, was quoted as saying that "overpopulation is perhaps the biggest problem facing

---

364 Khamsi, R., "DDT, Dutch elm disease & Dartmouth: Charles F. Wurster's crusade to make science matter," *Dartmouth Undergrad J of Sci*, vol IV, no. 1 (Fall 2001): 4–6, http://www.dartmouth.edu/~dujs/2001F/ddtdutch.pdf.

365 U.S. Supreme Court, *Murphy v. Butler*, 362 U.S. 929 (1960), http://bulk.resource.org/courts.gov/c/US/362/362.US.929.662.html (accessed Apr. 14, 2009). The first court case against DDT, to stop spraying of DDT in Long Island, New York, was initiated in 1957. The case made its way to the Supreme Court in 1960, but the Supreme Court refused to hear the case.

366 Natural Resources Defense Council, "U.S. law and policy," http://www.nrdc.org/legislation/helaw.asp (accessed Apr. 14, 2009).

367 Khamsi, R., "DDT, Dutch elm disease & Dartmouth: Charles F. Wurster's crusade to make science matter".

368 While the Sierra Club was formed in 1892, and so predates most environmental activist organizations, it definitely advanced as part of the environmental movement that came to life in the 1960s.

us,"[369] encouraged Paul Ehrlich to write a book on the problems of human population growth. Ehrlich published his bestseller, *The Population Bomb*, in 1968. The main theme of his book was that human population growth was the root cause of society's modern environmental problems.[370] Ehrlich conjured public fear by predicting dire scenarios of worldwide famines between 1970 and 1985 (none of which came true then or since).[371] Perhaps overlooked by many who read the book, Ehrlich also picked up Carson's anti-DDT theme. In a May 1970 issue of *Audubon*, Ehrlich even warned that DDT and other chlorinated hydrocarbons "may have substantially reduced the life expectancy of people born since 1945."[372] Fear tactics proved to be just as important in scaring people about population growth as they were in the war on pesticides. Ultimately, *The Population Bomb* promoted concerns that DDT caused rapid population growth as it reduced the burdens of malaria.

Ehrlich and David Brower were not alone in working against population growth and DDT. Attorney Dick Bower, Ehrlich, and Charles Remington formed the Zero Population Growth (ZPG) organization in 1968.[373] In Michigan, Dr. Lewis Batts, a medical doctor and bird-lover, worked to achieve a DDT ban. Like Wurster, Batts was one of the founders of the EDF. He was also

---

[369] *New York Times*, "Environmental leader quits Sierra board," May 20, 2000, http://www.nytimes.com/2000/05/20/us/environmental-leader-quits-sierra-board.html (accessed Apr. 14, 2009).

[370] Ehrlich, *The Population Bomb*, 1971.

[371] Michigan State University, "Founder of Zero Population Growth to speak at Michigan State's advanced degree ceremony," MSU press release, Jan. 13, 2007, http://newsroom.msu.edu/site/indexer/2743/content.htm (accessed Apr. 14, 2009).

[372] Bailey, R., "Earth Day, then and now," *Reason*, May 2000, http://reason.com/news/show/27702.html (accessed Apr. 14, 2009).

[373] Ryerson, B. "Visionary co-founder of population connection dies" *The Reporter*, vol. 39(2), Fall 2007.

a member of ZPG.[374] Batts pledged $10,000 to support legal action against DDT and another chlorinated hydrocarbon, dieldrin.

When the Michigan Department of Agriculture decided to spray dieldrin against the Japanese beetle, a highly destructive plant pest, DDT opponents, including the EDF, used the occasion as a pretense to carry out legal action against DDT. They achieved a statewide ban against most uses of DDT in April 1969.[375]

In Wisconsin, the court battle over DDT started December 2, 1968. Senator Gaylord Nelson was the opening speaker against the use of DDT. According to Hugh Iltis, a professor of biology and long-time supporter of Nelson, the hearings "dragged on for a year and [led to] the eventual banning of DDT in Wisconsin and four years later to victory in the banning of DDT nationwide."[376] Senator Nelson went on to be recognized as founder of Earth Day,[377] an event first held on April 22, 1970. He was also an avid believer that the major problem facing the world was uncontrolled growth in human populations. In his own words, "The same powerful forces which create the crisis of air pollution also are threatening our freshwater resources, our woods, our wildlife, and the scenic beauty of the nation. These forces are *the rapid increase in population*, industrialization,

---

[374] Zero Population Growth was an organization dedicated to reducing the rate of growth in human populations to zero. In other words, the rate of human births would be equal to the rate of human deaths.

[375] Michigan Environmental Council. "Lew Batts: Key player in Michigan's environmental turnaround", http://www.mecprotects.org/lewbatts.html (accessed Apr. 14, 2009).

[376] Iltis, H.H., "Gaylord Nelson: Fighter for the environment, defender of the Wild Ones, visionary of living within limits," http://www.for-wild.org/wchf/htm/GaylordNelsonIltis.htm (accessed Apr. 14, 2009).

[377] Webber, D.J., "Senator Gaylord Nelson, Founder of Earth Day," University of Missouri, 1996, http://web.missouri.edu/~polidjw/Nelson.html (accessed Apr. 14, 2009).

urbanization and *scientific technology*" (emphasis added).[378] In this statement, Nelson enunciates some of the main themes of environmentalism. One is to reduce population growth, and another is opposition to technology. Given DDT's association with both population growth and technology, it seems clear why the environmental movement would dedicate itself to DDT elimination. Though many environmentalists may have done a lot of good in exposing serious problems of pollution and endangered wildlife, there were some highly influential individuals within the movement that used their power and influence to campaign on population growth and specifically against DDT on those grounds.

The issue of withdrawing spray programs as a means of population control was broadly discussed and debated. Garrett Hardin, a leader in population control, believed that "every life saved this year in a poor country diminishes the quality of life for subsequent generations."[379] Likewise, in the prologue of *The Population Bomb*, Ehrlich announced with great authority, "In the 1970s and 1980s hundreds of millions of people will starve to death in spite of any crash programs embarked upon now. At this late date nothing can prevent a substantial increase in the world death rate."[380] We are now more than forty years past the publication date of Ehrlich's book. Hundreds of millions of people did not starve to death as Ehrlich predicted. Today, with a global population approaching seven billion, enough food is still being produced (although there are problems and inequities in food production, distribution, and sales).[381]

---

[378] Iltis, H.H., "Population prophet," FightingBob.com, Dec. 4, 2005, http://fightingbob.com/article.cfm?articleID=462 (accessed Apr. 15, 2009).

[379] Hardin, "Stalking the wild taboo."

[380] Ehrlich, *The Population Bomb*, xi.

[381] Leahy, S., "Population: Global food supply near the breaking point," Internet Press Service News Agency, Feb. 1, 2007, http://ipsnews.net/news.asp?idnews=33268.

At the end of his chapter on "the problem," Ehrlich concludes that

> there are only two kinds of solutions to the population problem. One is a "birth rate solution," in which we find ways to lower the birth rate. The other is a "death rate solution," in which ways to raise the death rate—war, famine, pestilence—*find us*. The problem could have been avoided by population control, in which mankind consciously adjusted the birth rate so that a "death rate solution" did not have to occur.[382]

The last sentence is written in the past tense, as if there is no longer a solution other than a "death rate solution." As background to this conclusion, Ehrlich argues that use of medical science in reducing death rates in developing countries contributes to the problem of population growth, stating that

> The development of medical science was the straw that broke the camel's back. While lowering death rates in the ODCs [overdeveloped countries] was due in part to other factors, there is no question that "instant death control," exported by the ODCs, has been responsible for the drastic lowering of death rates in the UDCs [underdeveloped countries].

As Ehrlich goes on to explain, the export of death control that he refers to is the use of DDT for malaria control.

> The introduction of DDT in 1946 brought rapid control over the mosquitoes which carry malaria. As a result, the death rate on the island [Ceylon] was halved in less than a decade. The death rate in Ceylon in 1945 was 22 [per 1000]. It dropped 34% between 1946 and 1947 and moved down to ten in 1954. Since the sharp postwar

[382] Ehrlich, *The Population Bomb*, 17.

> drop it has continued to decline and now stands at eight. Although part of the drop is doubtless due to the killing of other insects which carry disease and to other public health measures, most of it can be accounted for by the control of malaria.[383]

Most people would consider such a dramatic reduction in a country's death rate to be a marvelous outcome of an effort to reduce the disease and suffering of poor people. Most people could recognize that any population-growth problem is a separate problem, which should be dealt with separately from working against the use of DDT for control of malaria. That does not appear to be Ehrlich's point of view. He sheds more light on his perspective in comments about malaria and population control in the South American country of Colombia.

> Death control [DDT use] did not reach Colombia until after World War II. Before it arrived, a woman could expect to have two or three children survive to reproductive age if she went through ten pregnancies. Now, in spite of malnutrition, medical technology keeps seven or eight alive. Each child adds to the impossible financial burden of the family and to the despair of the mother.[384]

Ehrlich shows no insight into the despair of a mother or father from the loss of a child. Perhaps he has never known of a woman who has watched as all her children and husband die from malaria. Regardless, it was against this background of hysterical concern about growth in human populations that the environmental movement carried out its litigation and publicity wars against any and all uses of DDT.

---

[383] Ibid., 16.

[384] Ibid., 22.

Both *Silent Spring* and *The Population Bomb* criticized the use of DDT. Carson claimed that the justification for public-health use of DDT didn't make sense, that DDT quickly became ineffective and only made problems worse. Her basic thesis was any use of insecticide would select for resistance and the insecticide would lose its effectiveness. She claimed that it would select for "tough, resistant strains,"[385] so that more chemical would be required to get the same level of kill or else a more poisonous chemical would need to be developed. Carson was wrong on both claims. Resistance is not dealt with by using more of a public health insecticide. Furthermore, resistance signals a need for another mode of chemical action, not a more toxic chemical. Ehrlich, on the other hand, claimed that public-health use of DDT was so effective that it was an unacceptable contribution to limiting death, which in turn contributed to rapid population growth. Remarkably, the anti-DDT movement has been largely based on two entirely contradictory statements by Carson and Ehrlich. But contradictions aside, both books figure prominently in the creation of the modern environmental movement and, to this day, the two books stand as pillars of environmental theology.

If U.S. activism against DDT had stopped at U.S. borders, we might be inclined to assume that environmentalist motivations were directed at improving environmental conditions in the United States alone. However, as we will show later, and as exemplified in the international negotiations at the Stockholm Convention on Persistent Organic Pollutants, the environmental movement was hell-bent on eliminating DDT from malaria-control programs worldwide.

---

385 Carson, R, *Silent Spring* 1972, 237

Anti-DDT Litigation

The 1960s legal actions against DDT by environmental groups in New York, Wisconsin, Michigan, and elsewhere culminated in the 1972 EPA ruling banning DDT (see appendix 5 for more detailed accounts of the legal actions against DDT). These attempts to gain through the courts what could not be achieved through science were an exercise in emotive fear tactics and environmental politics. The most significant of these, the EPA's consolidated hearing, started in 1971 and continued until April 1972. Analyses of hearing records by Robert Ackerly, the chief trial counsel from the DDT industry, and Dr. J. Gordon Edwards, a highly respected professor of entomology at San Jose State University, show that key witnesses for the EPA and the EDF did not present credible testimony. As a result of that testimony, Hearing Examiner Edmund Sweeney filed his opinion, recommending that DDT not be banned:

> DDT is not a carcinogenic hazard to man. DDT is not a mutagenic or teratogenic hazard to man. The uses of DDT under the registration involved here do not have a deleterious effect on freshwater fish, estuarine organisms, wild birds or other wildlife. The adverse effect on beneficial animals from the use of DDT under the registrations involved here is not unreasonable on balance with its benefit. The use of DDT in the United States has declined rapidly since 1959. The Petitioners have met fully their burden of proof. There is a present need for the continued use of DDT for the essential uses defined in this case.[386]

The hearing examiner offered his opinion on the value of cross-examination:

[386] Sweeney, E.M., "EPA hearing examiner's recommendations and findings concerning DDT hearings," Apr. 25, 1972, 40 CFR 164.32.

> I think the right of cross-examination spurred a genuinely sober assessment of the facts available, particularly on the question of the benefits and risks of DDT; and it exposed those few instances where the purpose was to generate more heat than light on the subject.[387]

The judge also offered his opinion on the chemicals that might be considered as DDT replacements if DDT were banned:

> Although it was not in issue here, there was ample evidence to indicate that DDT is not the sole offender in the family of pesticides; and that necessary replacements would in many cases have more deleterious effects than the harm allegedly caused by DDT.[388]

The judge also commented on the credibility of the witnesses, noting that "there were some appalling instances of incredible inactions such as the publication of a paper containing faulty information which, after discovery, was never corrected and, apparently, is still being relied upon."[389] This appalling instance was a paper by George Woodwell published in *Science* magazine in 1967.[390]

Judge Sweeney presented his opinion in April 1972 after eight months of trial, "during which time the Examiner called 125 witnesses, entered 365 exhibits into the record and presided over a proceeding that produced a 9,312-page transcript. This was an extraordinarily thorough hearing."[391] Yet, two months

---

[387] Ackerly, R.L., "DDT: A re-evaluation. Part II," *Chemical Times and Trends*, 1982: 52.

[388] Ibid.

[389] Ibid.

[390] Ibid.

[391] Edmund Sweeney, Introduction to the Examiner's Report (1972). Sweeney said "no Hearing Examiner will ever enjoy the privilege that I had in listening to so many leaders in the field of scientific and medical

later, on June 2, 1972, the EPA's administrator, William D. Ruckelshaus, issued his opinion,[392] ignoring the results of the hearing and canceling all uses of DDT for crop production and non-health purposes in the United States, strongly implying in his opinion that DDT was almost assuredly toxic to humans.[393]

---

achievement . . . No restrictions were placed on the number of witnesses they could present, other than the necessary exhortations concerning relevance and materiality. The pros and cons of DDT have been well aired. I think the right of cross-examination spurred a genuinely sober assessment of the facts available, particularly on the question of the benefits and risks of DDT." EPA, Consolidated DDT Hearing, Hearing Examiner's Recommended Findings, Conclusions, and Orders (40 CFR 164.32) April 25, 1972, p16.

[392] Consolidated DDT Hearings. I.F. & R. Docket Nos. 63, et al. United States Environmental Protection Agency, Environmental Appeals Board. In The Matter of Stevens Industries, Inc., et al. Before the Administrator, U.S. Environmental Protection Agency; Opinion by William D. Ruckelshaus. June 2, 1972.

[393] 37 Fed. Reg. 13369 (July 7, 1971). Nixon was apparently furious about the decision to ban DDT. "I completely disagree with this decision," he wrote, and declared that he wanted "plenty of effort to get it reversed." Flippen, J Brooks, *Nixon and the Environment,* (Albuquerque, University of New Mexico Press, 2000) p. 172

Nixon clearly gets no points for consistency, having initially supported the moves to ban DDT.

WHO proved prescient in its fear that the U.S.' actions on DDT would affect world use. The EPA didn't think it would be a problem. Ruckelshaus's attitude concerning use of DDT and global public health was formed even before the consolidated hearings on DDT conducted by Edmund Sweeney and is revealed in a 1971 EPA document: "nonetheless, this Agency will not permit the triumphs of public health achieved in the past to be a continuing justification for use of a particular substance in future. To this extent, the requirements for use of economic poisons in a relatively developed country such as the United States may force a divergence from what is permitted in the developing countries where the public health impetus for control of such disease as malaria may require continuing use of pesticides whose side effects would no longer be tolerable here." Environmental Protection Agency "Reasons underlying the registration decisions

While the EPA reserved the use of DDT for emergencies, particularly public-health emergencies, this ban effectively ended the use of DDT in the United States and compromised its use in the rest of the world.[394]

Edwards described the administrator's lack of attention to the administrative hearing and the trial transcript:

> EPA Administrator William Ruckelshaus did not attend a single day of the seven months of EPA hearings on DDT, and aides reported that he did not even read the transcript (Santa Ana Register, 23 July 1972).[395]

Ruckelshaus's opinion was entirely contradictory to the scientific findings of seven months of testimony. For example, Ruckelshaus found that DDT presents a carcinogenic risk.[396] Based on animal-test data, he concluded that DDT "should be considered a potential carcinogen."[397] In contrast, Sweeney concluded that DDT is not a carcinogenic hazard to man.[398] On the subject of possible replacements for DDT, Sweeney concluded that leading replacement chemicals were more dangerous than DDT. Ruckelshaus's opinion addressed the issue of a replacement chemical differently. He recognized methyl parathion as the chemical that would be the primary DDT replacement, and he acknowledged that deaths had resulted from

---

concerning products containing DDT, 2,4,5-T, Aldrin and Dieldrin." March 18, 1971, *EPA*, Washington DC., 8

394 Other governments, especially European ones, had already banned the use of DDT.

395 Edwards, J Gordon "Pesticides in medicine and politics" Prepared for address to Doctors for Disaster Preparedness, San Diego, CA 14 June 1997. Copy on file with authors. p. 36

396 Ackerly, "DDT: A re-evaluation," 53.

397 Dunlap, T.R., *DDT: Scientists, citizens, and public policy* (Princeton, NJ, Princeton University Press, 1981), 234.

398 Ackerly, "DDT: A re-evaluation," 51.

operational use of methyl parathion. (In comparison, no human deaths or even illnesses had resulted from the operational use of DDT.) To allow for the increase in toxic risk from the use of methyl parathion, the EPA allowed a six-month waiting period before the full weight of the opinion would come into effect, meaning the order would not be effective until December 31, 1972. This waiting period was meant to allow time for the USDA and the EPA to provide training for operators of spraying equipment and others to use a much more dangerous insecticide.[399] This part of the ruling, more than any other aspect, shows how the EPA opinion was designed to hand a political victory to the environmental activists. At the time of the Ruckelshaus opinion, the EPA knew, from almost twenty-seven years of widespread DDT usage, that DDT was not known to cause human deaths or even human illness. In 1972, the EPA also knew, and openly admitted, that methyl parathion was a documented cause of human deaths.

In 1975, the EPA submitted an assessment to the U.S. House of Representatives of the scientific and economic aspects of its decision to delist DDT for use in agriculture. In its assessment of poisonings associated with accidental exposures to parathion and methyl parathion, they found that

> parathion and methyl parathion are the pesticides most frequently cited in incidents involving accidental exposure to pesticides. Preliminary data from the EPA Pesticide Accident Surveillance System (PASS) shows that parathion is the third and methyl parathion is the fifth most frequently cited pesticide in 1973. Based on an analysis of PASS data, Osmun (1974) stated that for 1972 and 1973, parathion and/or methylparathion were connected with 78% of the reported episodes relating to agricultural jobs, particularly those involving fields

[399] Consolidated DDT Hearings, 1972.

sprayed with pesticides for which safe reentry times for workers had been set.[400]

Not until twenty-seven years after promoting methyl parathion as a substitute for DDT did the EPA finally come to terms with the risks of methyl parathion. The agency accepted voluntary cancellation of many registered uses of methyl parathion in 1999 with an assessment that "methyl parathion is hazardous to workers—people who handle or apply the pesticide as part of their occupation, and people who work in fields to harvest treated crops. *Protective clothing and equipment are not sufficient to reduce the risks to workers to acceptable levels*" (emphasis added).[401] So, twenty-seven years after being forced to use methyl parathion, history has proven that Sweeney was right—DDT is not a human carcinogen, and the primary replacement insecticide was truly more dangerous than DDT. The EPA's tradeoff was clear: risk of poisoning and death for innocent Americans in exchange for a victory of environmental activism.

Until very recently, U.S. development policy completely ignored this risk–risk consideration, arguing that the United States can't support the use abroad of any substance that it doesn't use at home, even if the risks are completely different, and even if the substance is much safer to use than people think. ''For us to be buying and using in another country something we don't allow in our own country raises the specter of preferential treatment,'' said E. Anne Peterson, Assistant Administrator for Global Health at USAID. ''We certainly have to think about 'What would the American people think and want?' and 'What

---

[400] Environmental Protection Agency (EPA), "DDT, A review of scientific and economic aspects of the decision to ban its use as a pesticide." Prepared for: Committee on Appropriations, U.S. House of Representatives. Washington, DC, July 1975.

[401] EPA, "Methyl parathion risk management decision," Aug. 2, 1999, http://www.epa.gov/pesticides/factsheets/chemicals/mpfactsheet.htm (accessed Apr. 14, 2009).

would Africans think if we're going to do to them what we wouldn't do to our own people?'"[402] Many years after his decision to ban DDT, Ruckelshaus, in an interview with the *New York Times*, reported to be mystified by this position: "But if I were a decision maker in Sri Lanka, where the benefits from use outweigh the risks, I would decide differently. It's not up to us to balance risks and benefits for other people. There's arrogance in the idea that everybody's going to do what we do. We're not making these decisions for the rest of the world, are we?"[403]

Eliminating the Use of DDT Overseas

The actions of activists and environmental organizations did not end with the DDT ban in the United States. Immediately after the ban, the EDF appealed the decision, seeking to end even the production and export of the banned chemical.[404] This appeal shows the environmental movement's determination to stop all uses of DDT, even those uses that protected human health. As we describe in chapter 7, this determination continued into the 21st century when the UN Environment Program's Stockholm Convention on Persistent Organic Pollutants (POPs) provided a further opportunity to stop public health use of DDT.

Environmentalists' determination and the successes of their legal actions against DDT paved the way for the growth of the movement. In a short span of time, the activist organizations grew from grassroots to extremely well-paid activism. A 1977 article on the environmental complex reported that the twelve leading environmental organizations combined had over $48 million in revenues.[405] In 2002, the Natural Resources Defense

---

[402] Rosenberg, "What the World Needs Now Is DDT."

[403] Ibid.

[404] Dunlap, *DDT*, 234.

[405] The Heritage Foundation, "The environmental complex," Nov. 19, 1977,

Council (NRDC) and the EDF (now known as Environmental Defense) each had annual revenues of over $40 million.[406] Edwards noted that "Ruckelshaus, alleging adverse effects of DDT, signed fund-raising letters on behalf of the Environmental Defense Fund. On his personal stationery, he wrote, 'EDF's scientists blew the whistle on DDT by showing it to be a cancer hazard, and three years later, when the dust had cleared, EDF had won.'"[407]

While the EDF had Ruckelshaus, the NRDC also had a powerful supporter: George M. Woodwell, a leader in the anti-science/technology and anti-population-growth movement. He advocated for restricting technology to make its use more and more expensive and advocated against growth of human populations. In *Science* (April 1970), he wrote that these issues were important in solving the problems of environmental pollution. Woodwell wrote, "The solution? Fewer people, unpopular but increasing restrictions on technology (making it more and more expensive)."[408] Although Woodwell was among those whose testimony was largely discredited in the EPA hearing, his role in environmental activism remained well established. With Woodwell as vice chairman of the NDRC's board of trustees,[409] which directed the group's activities, the NRDC started to push the DDT fight outward from the borders of the United States. In 1977, the Audubon Society and NRDC filed a suit against USAID to stop the export of DDT. As a

http://www.heritage.org/Research/EnergyandEnvironment/IA4.cfm (accessed Apr. 14, 2009).

[406] Arnold, R., "The politics of environmental defense," Aug. 2004, Capital Research Center.

[407] Sirkin, G. and Sirkin, N., "DDT, fraud, and tragedy," *The American Spectator*, Feb. 25, 2005, http://www.spectator.org/dsp_article.asp?art_id=7812 (accessed Apr. 14, 2009).

[408] Woodwell, GM., "Effects of pollution on the structure and physiology of ecosystems." *Science*, 20 April 1970, 168:429-433

[409] The Heritage Foundation, "The environmental complex."

result, USAID "undertook to discourage other countries from using DDT by threatening to stop foreign aid to any country using it. Its threat spread Ruckelshaus's ban worldwide."[410]

Over time, through misinformation and political propaganda, the campaign to stop what was described by Paul Ehrlich as the "export of death control" succeeded.[411] DDT was largely removed from disease-control programs, and preventable diseases and deaths increased. Infectious disease statistics for malaria, dengue, leishmaniasis, and bartonellosis (an insect-borne bacterial disease) from endemic countries around the world confirm the tragic results of anti-DDT activism. Those within the environmental movement who wanted to use malaria as a method of limiting growth of human populations succeeded in largely eliminating effective disease-control programs, and use of DDT in particular.

DDT's Safety Record Persists

More than thirty years after the EPA banned DDT, it still retains an impressive record of safety for humans. Regardless, the controversy over DDT and human health persists. One of the longest-raging debates is about whether DDT causes cancer. Proponents of the DDT–cancer connection are quick to point out that WHO has designated DDT a possible human carcinogen, but years of research and epidemiological studies have failed, in a uniform and consistent manner, to link DDT to cancer in humans.[412]

---

[410] Sirkin and Sirkin, "DDT, fraud, and tragedy."

[411] Ehrlich, *The Population Bomb*, 1971.

[412] "Not one case-control study of DDT's human carcinogenicity has been affirmatively replicated." DeGregori, T., *Bountiful Harvest* (Washington, DC: The Cato Institute, 2002), 131–36;

Agency for Toxic Substances and Disease Registry, "Toxicological profile for DDT/DDD/DDE," http://www.atsdr.cdc.gov/toxprofiles/tp35.html (accessed Apr. 14, 2009).

One cancer to which researchers repeatedly tried to link DDT was breast cancer, but several studies have refuted this link.[413] One researcher, David Hunter, put the lack of correlation quite bluntly:

> We know that breast cancer is at least in part a hormonally-determined disease and the prevailing wisdom is that a woman's lifetime exposure to estrogen is a major determinant of lifetime breast cancer risk. Since organochlorines have been implicated as endocrine disruptors in so many different systems, it makes sense to hypothesize that these chemicals could also influence the rate in women of endocrine and estrogen related diseases such as breast cancer . . . The field really started when some early small epidemiological studies suggested a positive relation between some sentinel, widely used organochlorine chemicals . . . A lot of studies have been done since and . . . they have all been essentially null with respect to DDE in adipose tissue. A large European study, in fact, observed a statistically significant inverse trend: the higher the level of DDE, the lower the risk of breast cancer. Some large, more recent U.S. studies have essentially observed no association. A large study in Canada similarly saw no association. So these studies [of] DDE in adipose tissue certainly are not indicating

[413] Calle, E., et al., "Organochlorines and Breast Cancer Risk," *CA Cancer J Clin,* vol. 52, no. 5 (Sep–Oct 2002): 301–309, http://caonline.amcancersoc.org/cgi/content/short/52/5/301 (accessed Sept. 25, 2009), Hunter, D., et al., "Plasma Organochlorine Levels and the Risk of Breast Cancer," *N Engl J Med*, vol. 337, no. 18 (Oct. 1997): 1253–58; Safe, S., Editorial: "Xenoestrogens and Breast Cancer," *N Engl J Med*, vol. 337, no. 18 (Oct. 1997): 1303–04.

any positive association between DDE, the major metabolite of DDT, and the risk of breast cancer.[414]

Despite a lack of evidence to support the link between DDT and cancer in humans, those people who are understandably desperate to find a cause for breast cancer refuse to let go of their preconceptions and accept the data. When told that the Hunter study showed no link between DDT and breast cancer, Geri Barish, president of 1 in 9: The Long Island Breast Cancer Action Coalition, said, "I just find it very difficult to believe. I can't accept it at all."[415]

Attempts have also been made to link DDT to pancreatic cancer, Hodgkin's disease, non-Hodgkin's lymphoma, multiple myeloma, prostate and testicular cancer, endometrial cancer, and general cancer incidence. Yet, despite sixty years of research and hyperbolic claims, the Agency for Toxic Substances and Disease Registry states that "the existing information does not support the hypothesis that exposure to DDT/DDE/DDD increases the risk of cancer in humans."[416] Furthermore, studies of exposure to DDT in humans have found no association between DDT and generally higher mortality rates or

[414] Hunter, D., "1,1-Dichloro-2,2-Bis (P-Chlorophenyl) ethylene and polychlorinated biphenyls and breast cancer: combined analysis of five U.S. studies," Harvard Medical School and School of Public Health, Presentation at International Symposium on Environmental Endocrine Disruptors, Dec. 2001, Ministry of the Environment, Japan, http://www.env.go.jp/chemi/end/2001report/pdf_e/hunter_e.pdf. (accessed Sept. 25, 2009).

[415] Kolata, G., "Study discounts DDT role in breast cancer," *New York Times*, Oct. 30, 1997, http://www.nytimes.com/1997/10/30/us/study-discounts-ddt-role-in-breast-cancer.html?scp=1&sq=Study%20 Discounts%20DDT%20Role%20in%20Breast%20Cancer&st=cse (accessed Apr. 14, 2009).

[416] Agency for Toxic Substances and Disease Registry, Toxicological profile for DDT, DDE, DDD, 2002, http://www.atsdr.cdc.gov/toxprofiles/tp35.html, 107-108, (accessed Sept. 25, 2009).

respiratory, cardiovascular, gastrointestinal, hematological, musculoskeletal, hepatic, dermal, or ocular effects.[417]

DDT and Babies

In 2001, scientists thought they had found proof of DDT's possible function as an endocrine disruptor. "DDT Finally Linked to Human Health Problems" ran the headline in the *New Scientist* when a 2001 study linked pre-term deliveries to DDT exposure.[418] The relief among anti-DDT campaigners was palpable; it had taken sixty years, but finally something detrimental could be laid at DDT's door.

The study found a link between DDE levels in mothers' blood and pre-term births. Before this study, others tried to establish a link between DDT and babies with low birth weights, but these studies were inconclusive.[419] The author of the pre-term birth study, researcher Matthew Longnecker of the U.S. National Institute of Environmental Health Sciences in North Carolina, is quite certain as to the significance of this study. "One of the reasons this finding is important is there are not any generally accepted adverse health effects of exposure to DDT or its metabolite, DDE, in humans," he said.[420] "If indeed—and it's a big 'if'—the premature births found in the analysis were a direct result of DDT exposure, and if those premature babies had

---

417 Ibid., 74–84.

418 Longnecker, M.P., et al., "Association between maternal serum concentration of the DDT metabolite DDE and preterm and small-for-gestational-age babies at birth," *The Lancet*, vol. 358, no. 9726 (2001): 110–14.

419 Attaran A., et al., "Balancing risks on the backs of the poor," *Nature Medicine*, vol. 6, no. 7 (2000): 730.

420 Young, E., "DDT finally linked to human health problems," *New Scientist*, July 13, 2001, http://www.newscientist.com/article/dn1012-ddt-finally-linked-to-human-health-problems.html (accessed Apr. 14, 2001).

the same likelihood of survival as babies born prematurely for other reasons, as much as 15 per cent of infant mortality between 1959 and 1966 may have been due to DDT," continued Longnecker.[421] "In earlier decades in the U.S., we may have had an epidemic of pre-term births that we are just now discovering . . . We have to be concerned about what might be happening in those 25 countries where DDT is still used."[422]

In another recent paper on DDT, Aimin Chen and Walter Rogan "estimate an increase in infant deaths [due to DDT] that is of the same order of magnitude as that from eliminating infant malaria."[423] Collectively, the paper by Longnecker and the paper by Chen and Rogan present a question as to the benefits of DDT for malaria control if, as they speculate, it is deleterious to infant survival. Pierluigi Cocco, a scientist specializing in occupational medicine and epidemiology, conducted a study along with several colleagues to try to answer this question.[424] Taking advantage of an intensive campaign of DDT spraying for malaria control in Sardinia, Italy, from 1946 to 1950, the researchers compiled and analyzed data on reproductive health before and during the DDT-spray campaign for both Sardinia and the whole of Italy. They found that the "stillbirth rate, infant mortality rate, and male/female ratio in newborns were apparently unaffected following widespread but focused use of DDT in Sardinia, Italy."[425] The authors concluded, "we speculate that predictive models of the reproductive outcomes in

---

[421] *Nature*, "Association between maternal serum concentration of the DDT metabolite DDE and preterm and small-for-gestational babies at birth." *Nature News Service*, vol. 10, no. 1038 (July 13, 2001), http://www.nature.com/nsu/010719/010719-3.html, (accessed Sept. 25, 2009).

[422] Young, "DDT finally linked to human health problems."

[423] Chen, A. and Rogan, W. "Non malarial infant death and DDT use for malaria control," *Emerg Inf Dis*, vol. 9 (2003): 960–64.

[424] Cocco, P., et al., "Reproductive outcomes following environmental exposure to DDT," *Repro Tox*, vol. 22 (2006): 5–7.

[425] Ibid., 5.

relation to DDT use for malaria control should include a likely side-effect of a successful anti-malarial program, such as the improvement in the maternal health status and nutrition."[426]

In other words, the authors suggest that benefits of malaria control might wash out any adverse effects of DDT exposure, even if there were adverse effects—adverse effects that remain unproven. While Cocco's research does not provide a complete answer to the question, his comments allude to the certainty that without DDT in malarious countries, women are giving birth to underweight and premature babies, and the effects are due to malaria, not DDT. We can be certain of this because pregnancy compromises a woman's immune system and pregnant women, even those who have previously acquired partial immunity to malaria, become susceptible to the disease again, often becoming anemic (anemia is often a result of infection because malaria parasites destroy red blood cells), which can be harmful to the developing fetus. Compromised or undeveloped immune systems explain, in part, why pregnant women and children under the age of five make up 90 percent of malaria deaths. If a pregnant woman survives a bout of malaria, she is, according to WHO, likely to suffer a range of adverse pregnancy outcomes, including low birth weight, spontaneous abortion, and neonatal death.[427]

Scientists within the malaria-control community also responded to Chen and Rogan's claims. They pointed out that "The available evidence . . . does not support the idea that the reported increase in premature births was a side effect of DDT use. In any case, the increase in deaths attributable to premature births was orders of magnitude less than the reduction in deaths directly caused by malaria and other conditions indirectly related to malaria." Similarly, one of this book's authors (Roberts)

[426] Ibid., 7.

[427] WHO, "Chapter 4: Malaria during pregnancy," in "Africa malaria report 2003," Geneva, http://rbm.who.int/amd2003/amr2003/amr_toc.htm, (accessed Sept. 25, 2009) .

observed in 2004 that "major benefits of DDT use were seen in Guyana, where in 2 to 3 years, near elimination of malaria halved maternal deaths and reduced infant deaths by 39%. Anemia-associated deaths in pregnant females were reduced from 10 to 2.3 per 1,000 adult deaths. There was no offset of infant deaths attributable to adverse effects of DDT."[428] Longnecker has subsequently changed his position regarding the effect of DDT on human health. In 2005, he said that "the data available on the potential health effects of DDT and its degradation products suggest that there may be no serious consequences of exposure, though some associations that have been reported require further study."[429]

So what effect does DDT have on humans? The preponderance of sixty years of evidence agrees with the opinion of the National Academy of Sciences: "*there is still no clinical or epidemiological evidence of damage to man from approved uses of DDT*" (emphasis added).[430] A. G. Smith of the Medical

[428] Roberts D.R., et al., "Malaria control and public health (replies)," *Emerg Inf Dis*, vol. 10, no. 6 (June 2004), http://www.cdc.gov/ncidod/Eid/vol10no6/03-0787_03-1116.htm (accessed Apr. 14, 2009).

[429] Longnecker, M.P. Invired Commentary: Why DDT Matters Now. *AM J Epidemiol* 162 (8) (2005): 726-8.

[430] Committee on the Applications of Ecological Theory to Environmental Problems, *Ecological Knowledge and Environmental Problem Solving: Concepts and Case Studies* (Washington, DC: National Academies of Science Press, 1986): 364; "DDT's record speaks for itself. For years most Americans and Europeans ingested substantial amounts of DDT in food every day; the United States alone sprayed 70,000 tons on crops every year for 20 years. Kids on bikes weaved in and out of the DDT clouds blown over the streets in countless American towns to control mosquitoes. Many millions of homes in Asia, Southern Africa and Latin America have been sprayed once or twice a year with DDT. Yet no adverse health effects have been reported. 'Scientists have searched exhaustively, but found nothing substantial,' says Mary Galinski, a molecular biologist at Emory University in Atlanta, Georgia, and president of Malaria Foundation International." Dyson, J., "DDT Should Not Be Banned,"

Research Council's Toxicology Unit at the UK's University of Leicester also summed up the prevailing opinion of DDT toxicity in a letter in *Lancet*:

> The early toxicological information on DDT was reassuring; it seemed that acute risks to health were small. If the huge amounts of DDT used are taken into account, the safety record for human beings is extremely good. In the 1940s many people were deliberately exposed to high concentrations of DDT through dusting programs or impregnation of clothes, without any apparent ill effect. There are probably few other chemicals that have been studied in as much depth as has DDT, experimentally or in human beings . . . In summary, DDT can cause toxicological effects but the effects on human beings at likely exposure are very slight.[431]

Why is DDT so surprisingly non-toxic? According to studies conducted by South African scientists H.E. Smit, Henk Bouwman, and David le Sueur in 1992, the skin poorly absorbs DDT in powder form, and this accounts for the safe handling record of DDT.[432] Studies have recorded that those living in households treated with DDT have higher levels of DDT in their

---

Reader's Digest (South Africa), Dec. 2000; "DDT is not toxic to humans and is the only commercial insecticide in widespread use which has not caused a human fatality through proper use." Leitch, R., "New Era, Old Disease," *U.S. Medicine*, Feb. 2003.

[431] Smith, A.G., "How toxic is DDT?" *Lancet*, vol. 36, no. 9226 (July 22, 2000): 267–68, http://www.malaria.org/smithddt.html (accessed Apr. 14, 2009).

[432] Smit, H.E., et al. "DDT and community health," *J Compre Hlth*, vol. 3, no. 1 (1992): 175–78.

systems,[433] but while recording heightened levels of DDT in humans is one thing, proving that this leads to ill health is another. The small amount of DDT needed for malaria control,[434] along with the careful way in which the malaria programs use DDT, led Bouwman and others to observe no significant variations in the levels of DDT in the environment before and after the spraying of DDT for malaria control.[435]

The search for some evidence that DDT and DDE cause human-health harm continues in earnest. In May 2009, the journal *Environmental Health Perspectives* (EHP) published "The Pine River Statement: Human Health Consequences of DDT Use."[436] This statement was prepared by fifteen scientists from the United States and South Africa and reviews almost five hundred scientific studies on DDT/DDE and human health found in a PubMed (a medical literature database) search. The Pine River Statement discusses a wide range of potential human-health effects of DDT, including DDT as a cause of cancer, diabetes, fetal underdevelopment, shortened duration of

[433] Bouwman, H., et al. "Levels of DDT and metabolites in breast milk from KwaZulu mothers after DDT application for malaria control," *Bull Wrld Hlth Org*, vol. 68 (1990): 761–68.

[434] "Not only did government regulators minimize scientific evidence of the safety and effectiveness of DDT, they also failed to appreciate the distinction between its large-scale use in agriculture and more limited application for controlling carriers of human disease. Although DDT can be a toxic substance, there is a big difference between applying large amounts of it in the environment—as American farmers did before it was banned—and applying it carefully and sparingly to fight mosquitoes and other disease-carrying insects. A basic principle of toxicology is that the dose makes the poison." Miller, H.I., "Bring Back DDT," *New York Times*, Aug. 7, 2003.

[435] Bouwman, "Levels of DDT and metabolites in breast milk."

[436] Eskenazi, B., Chevrier, J., Rosas, L.G.,et al., "The Pine River statement: human health consequences of DDT use," *Environ Health Persp*, doi: 10.1289/ehp.11748, May 4, 2009, http://www.ehponline.org/members/2009/11748/11748.pdf, (accessed Sept. 25, 2009).

lactation, reduced child growth, reproductive problems, and neurodevelopmental problems.

The Pine River Statement is one of many assessments investigating potential human-health harm caused by DDT. And in line with most of these assessments, the Pine River Statement's human-health case against DDT is weak. In order to prove a cause-and-effect relationship between DDT and human-health harm, certain core criteria should be met, such as strength of association, biologic credibility, and consistency with other investigations. Two other important, although sometimes considered weaker, criteria are time sequence (cause must precede effect) and a proportional dose–response relationship. None of the studies presented in the Pine River Statement satisfy these criteria. The studies presented are un-replicated, contradictory, or statistically insignificant.

Yet, despite the lack of any credible evidence, the authors conclude that they present evidence that DDT may pose a risk to human health and go on to argue for further scientific studies with a particular focus on human exposure to DDT when used in malaria control. One has to wonder when the endless investigations will end, the millions of dollars it costs to conduct studies into DDT's supposed links to human-health harm will stop being wasted, and when public-health authorities and governments will start to fund studies in the search for new public health insecticides.

A study published in the British Journal of Urology International in November 2009 purported to find an association between DDT house spraying for malaria control and urogenital birth defects (UGBD) in boys.[437] The study concluded that IRS would increase the probability that a boy would be born with a UGBD by 33%. This claim would be worrying were it not for the fact that it is only arrived at after manipulating data,

[437] Bornman R, et al "DDT and urogenital malformations in newborn boys in a malarial area" *Brit. J. Urol. Int*, doi:10.1111/j.1464-410X.2009.09003.x, Nov. 2009.

lowering the threshold of statistical significance and allowing serious shortcomings in the study design and methodology.[438] Despite these problems and the fact that these UGBDs have not been found in any other area where DDT was used in great quantities, the study was widely covered by the media in South Africa and elsewhere, giving the authors a powerful platform from which to lobby against DDT use in malaria control.

The evidence against DDT is exceedingly weak, and even if DDT were more toxic than it is, the potential risks would still have to be weighed against the very real threat of dying from malaria. As it stands, the low toxicity of DDT and the high risk of malarial infection indicate that public-health authorities should continue to use DDT in malarial control. In fact, its use should not have been stopped in the first place. After commissioning experts to review recent studies that suggested "an association between DDT and human cancers and a report on the presence of DDT in breast milk," the WHO Study Group on Vector Control for Malaria and Other Mosquito-Borne Diseases determined that "the information presented does not provide convincing evidence of adverse effects of DDT exposure as a result of indoor residual spraying as carried out in malaria control activities. There is therefore, at this stage, no justification on toxicological or epidemiological grounds for changing current policy towards indoor spraying of DDT for vector-borne disease control."[439]

[438] Africa Fighting Malaria, "A response to the paper entitled: DDT and urogenital malformations in newborn boys in a malarial area", 26 October, 2009, available from http://www.fightingmalaria.org/article.aspx?id=1353 (accessed Nov. 16, 2009)

[439] WHO, "Vector control for malaria and other mosquito-borne diseases," WHO Technical Report Series, no. 857, 1995.

The Popular Press, Science Journals, and DDT

So how has the environmental movement maintained its pressure against DDT, including the public-health use of DDT, and achieved such a remarkable victory over common sense and good science? The short answer is that many editors of science journals, some unwittingly, others intentionally, have aided the tactics of the environmental movement (and continue to do so), as have science writers in the popular press. There is no better example of a science journal fully subscribing to and, in a highly biased manner, aiding the environmental movement than what is seen in pages of *Science*. From 1968 to 1972, *Science*, widely considered to be the premiere scientific journal in the United States, published seventy mostly antagonistic papers, letters, and editorials on DDT.[440] Malaria-control scientists and others wrote letters to protest the many falsehoods in *Science* articles, but to no avail.[441]

A PubMed search by author reveals writings by leaders in the campaign against DDT in *Science* sixteen times for Paul Ehrlich, eight for George Woodwell, and eleven for Charles Wurster. A PubMed search for the keywords DDT + malaria + *Science* magazine will produce only one result: a letter to the editor.[442] That letter was written by Donald Roberts, one of this book's authors, and it was only reluctantly accepted and published in 1982. Of the DDT papers in *Science*, with the

---

[440] Unsupported claim by C.F. Wurster in 1968 that DDT was endangering the Bermuda petrel: Wurster, C.F. and Wingate, D.B., "Residues and declining reproduction in the Bermuda petrel," *Science*, vol. 159 (1968): 979–81; Unsupported claim by C.F. Wurster in 1968 that DDT killed algae: Wurster, C.F., "DDT reduces photosynthesis by marine plankton," *Science*, vol. 159 (1968): 1474–75.

[441] Russell, P.F. and Epstein, S.S., "DDT toxicity," *Science*, vol. 177, no. 47 (1972): 387–88; McLean, L.A., "DDT residues and Bermuda Petrels," *Science*, vol. 161, no. 3839 (1968): 397.

[442] Roberts, D.R., "Health problems of colonists," *Science*, vol. 217, no. 4559 (Aug. 6, 1982): 484.

exception of the 1982 letter and the letter by Paul Russell, a leading U.S. malaria scientist in the 1950s and 60s, no other paper published in *Science* suggests that DDT should be used for malaria control, that it is the most cost-effective insecticide available, that DDT functions as a repellent and irritant, or that DDT is safe for human exposure.

This record supports Edwards' claim that *Science* refused to publish articles favorable to DDT: "Abelson [the editor of *Science*] told Dr. Thomas Jukes that *Science* would never publish any manuscript by a protagonist of DDT."[443] Edwards attended the EPA hearings on DDT, and as a result of his experiences there and his inspections of literature from leaders of the environmental campaign, he became an ardent campaigner for truth in science. In 2002, he wrote to one of this book's authors (Roberts) with a paper he had prepared and submitted to *Science*. His intent was to expose misrepresentations of science that had occurred during the EPA hearings against DDT.[444] His paper attacked the anti-DDT bias of *Science* magazine and, perhaps as a result, his report was never published. He explained the basis for his paper as follows:

> After 40 years of concern resulting from untruthful statements and unsubstantiated allegations published in science journals, some desire for "truth in science" may now be emerging. An article in *Science* (7 June 2002) referred to scientific misconduct in Germany. There have also been abundant examples of deliberate fraud in American science. Much of the fraud has been environmental, but lies about chemicals have been widespread, and have adversely affected every American. The most direct threat to the health and

---

[443] Edwards, J.G. "Science, pesticides and environmentalist politics" Executive Intelligence Review, December 10, 1000 pp 20-31. (page 16).

[444] Edwards, J.G., written personal communication to Don Roberts.

> welfare of Americans is the propaganda regarding chemicals in the environment. The most widely attacked chemical, DDT, saved hundreds of millions of human lives from diseases transmitted by blood-sucking insects. Environmental scientists have complained that those lives should not have been saved, because "humans are far too abundant."[445]

Evidence that Dr. Edwards was correct in attributing bias to the editors of *Science* magazine in coverage of DDT issues is found in the history of papers the editors chose to publish. Charles Wurster figures prominently in this record of publication bias. Wurster was the chief scientist for the Environmental Defense Fund, and the EDF had become the lead organization in the fight against DDT. In this endeavor, the EDF seemed to have a powerful ally in the editors of *Science* magazine. From 1968 to 1972, Wurster, wrote ten percent of the DDT papers in *Science*, and most were scientifically weak. His paper on DDT endangering the Bermuda petrel was false.[446] He coauthored a paper with Woodwell in 1967 on DDT in an estuary.[447] During cross-examination in EPA hearings, Woodwell admitted that their estimated average of thirteen pounds of DDT per acre rested on highly biased sampling. The samples were taken from soil where the DDT-spray truck was washed.[448] Wurster's 1968

---

445 Edwards, J.G., "DDT: A case study in scientific fraud," *J Amer Phys & Surgeons*, vol. 9, no. 3 (2004): 83–88.

446 Wurster and Wingate, "Residues and declining reproduction in the Bermuda petrel."

447 Woodwell, G.M., et al., "DDT residues in an East Coast estuary: A case of biological concentration of a persistent insecticide," *Science*, vol. 156 (1967): 821–24.

448 In a 1967 *Scientific American* paper, Woodwell stated that "DDT residues in the upper layer of mud in this marsh ranged up to 32 pounds per acre." Ehrlich's *The Population Bomb* trumpeted only their maximum value of 32 pounds per acre. Woodwell, G.M., "Toxic

paper on DDT reducing photosynthesis by marine phytoplankton was nonsense.[449] As we explain in a footnote below, his test conditions were unrealistic and irrelevant in respect to what would occur in nature.[450]

Science editor Phil Abelson wrote four editorials about DDT and environmental pollutants: one in 1969, one in 1970 and two in 1971. In 1969, he wrote, "DDT and its relatives are not truly persistent but are slowly destroyed in soil. DDT is slowly degraded in man, and it is also excreted, so that concentrations do not build up indefinitely."[451] Yet despite Abelson's correct

---

substances and ecological cycles," *Scientific American*, vol. 216, no. 3 (1967): 24–31.

449 Wurster, C.F., "DDT reduces photosynthesis by marine phytoplankton," *Science*, vol. 159 (Mar. 29, 1968): 1474–75.

450 Wurster admitted that DDT solubility is roughly 1.2 parts per billion (ppb). In spite of this, he dosed his test vats with concentrations of up to 500 ppb—in other words, his vats were supersaturated (saturated to the point that no more chemical would go into solution) with DDT. He admitted that DDT quickly disappeared from such solutions, as it always does. His claims never dealt with the reality that DDT in suspensions of particulates or DDT in solids do not change the solubility factor of DDT in water. In other words, the amount of DDT in solids will not change the amount of DDT in water solution (1.2 ppb), and only the DDT in solution will be readily available to phytoplankton. His experiments showed no adverse effects on photosynthesis within the range of DDT solubility and, in fact, there was an uptake of carbon above 100 percent, as he described it, "at low concentrations of DDT." The low concentrations were actually those within the range of DDT saturation in water. In a repetitive cycle of one promoting the work of the other, his claim of DDT harming phytoplankton was picked up and trumpeted loudly by Ehrlich in *The Population Bomb*, where Ehrlich wrote that "DDT affects some kinds of planktonic plants more than others. This could lead to large changes in the plant plankton communities which are the basic source of energy for marine life. The results for our fisheries could be catastrophic." Ehrlich, *The Population Bomb*, 36.

451 Abelson, P.M., "Chemicals & Cancer," *Science*, vol. 166, no. 3906 (Nov 7, 1969): 633.

analysis, in 1970, *Science* published a major paper that included a model based on an assumption of no DDT degradation or excretion.[452] One of the authors of this paper was Victor Yannacone, the lawyer who directed the DDT lawsuits in Michigan and Wisconsin. The model predicted that concentrations of DDT in long-lived species would continue to build up long after additions of DDT to the environment ceased. The authors went on to predict possible systematic obliterations of species in higher trophic levels (levels of the food chain). At the time, in addition to Abelson's 1969 statement, several papers on degradation and excretion of DDT had already been published in *Science*.[453] Today, we know that DDT concentrations declined in all trophic levels after DDT was delisted for use in agriculture. We also know that there was no loss of species at all. In fact, there are not even any claims of species losses, and there were certainly no obliterations as suggested in the paper.

The 1970 *Science* paper by Yannacone and others was written as a major contribution to the campaign against DDT.[454] Another model paper, written by Woodwell, was published by *Science* in 1971 on the distribution of DDT in the biosphere.[455] Even though this paper showed that DDT "was degraded to innocuousness or sequestered in places where it is not freely

---

[452] Harrison, H.L., et al., "Systems studies of DDT transport," *Science*, vol. 170, no. 3957: 503–508.

[453] Mendel, J.L., and Walton, M.S., "Conversion of p,p'-DDT to p,p'DDD by intestinal flora of the rat", *Science* March 25, 1966, 151(717):1527-1528; and Jonnson, B.T, et al. "Conversion of DDT to DDD by pathogenic and saprophytic bacteria associated with plants" *Science*, August 4, 1967, 157(788):560-561; and Guenzi, W.D and Beard, W.E., "Anaerobic biodegradation of DDT to DDD in soil", *Science*, May 26, 1967, 156(3778):1116-1117

[454] Ibid.

[455] Woodwell, G., et al., "DDT in the biosphere. Where does it go?" *Science*, vol. 174, no. 4014 (Dec. 10, 1971): 1101–1107.

available to the biota," it was nevertheless appropriately attacked for "many items of speculation, suggestion, and opinion."[456]

Editors of *Science* continue in this vein with a policy forum article in 2005 authored jointly by Paul Ehrlich and Donald Kennedy (the present editor of *Science*). In this article, the authors seem to separate out the ideas of what must be, in their view, an elite set of scientists of the environmental movement.

> The steps [to an improved environment] that most members of the *relevant* scientific community believe are necessary (e.g., reduction of human-caused greenhouse gas emissions, establishment of marine reserves, limiting human population growth and per capita consumption) are disconnected from those measures the rest of society, and especially politicians, are willing to undertake (emphasis added).[457]

One must ask whether the "relevant scientific community" will proceed with environmentalist action on the basis of science, or, as mentioned here, in what they "believe." One also must wonder whether the same aggressive tactics (misinformation and fear) will be employed as have been employed in the past.

A similarly worrisome bias is common with science reporting in the popular press. The original small study by Mary Wolff suggesting a link between DDE and breast cancer garnered national headlines in 1993.[458] It was followed by half a

---

456 Stewart, C.A., "Atmospheric circulation of DDT," *Science*, vol. 177: 724–25.

457 Ehrlich, P.R., and Kennedy, D., "Millennium assessment of human behavior," *Science*, vol. 309 (July 22, 2005): 562–63.

458 Wolff, M.S., et al., "Blood levels of organochlorine residues and risk of breast cancer," *J Nat Canc Inst*, vol. 85, no. 8 (1993): 648–52; Associated Press, "Study ties DDT to cancer cases," Published in the *New York Times*, Apr. 22, 1993, p. A16; Raeburn, P. (Associated Press), "DDT exposure linked to breast cancer in women," *Seattle Post-Intelligencer*, Apr. 21, 1993, p. A3.

page of coverage on DDT, estrogen exposure, and breast cancer in the Forum section of the *Washington Post* on January 25, 1994.[459] Yet, a paper in 2001 reporting on many DDE-breast cancer studies that concluded DDE was not associated with breast cancer was reported in the last sentence of a small article with the title "Hormones may lower risk of breast cancer's return."[460] This article was buried on page A10 of the *Washington Post* and the last sentence stated, "In another study in the Journal [of the National Cancer Institute], no strong link was found between breast cancer and two chemicals, PCBs and DDE." The statement "no strong link was found" was wrong—the actual conclusion of the paper was, "combined evidence does not support an association" between DDE and breast cancer.[461] In other words, there was no link at all. So the public learned from newspapers in 1993 that DDE caused breast cancer, but they were essentially not informed, years later, that earlier reports were wrong and DDE was not a cause of breast cancer.

A systematic bias against DDT has led to a seriously misinformed public and science community. Suppose a large, population-based study demonstrated no apparent link between sizable occupational DDT exposure and cancer. These findings would be useful for fellow scientists to know and for authorities to bear in mind when making policy decisions. But such results are not exciting, and thus the chance of the paper being published in a top journal like *Science* is remote. As an example, consider a paper by notable toxicologists William Deichmann and William MacDonald showing that rates of liver cancer

---

459 "Breast Cancer's link to estrogen exposure," *Washington Post*, Forum section, Jan. 25, 1994, p. 13.

460 Laden, F., et al., "1, 1-dichloro-2,2-bis(p-chlorophenyl)ethylene and polychlorinated biphenyls and breast cancer: combined analysis of five U.S. studies," *J Nat Canc Inst*, vol. 93, no. 10 (2001): 768–76; Associated Press, "Hormones may lower risk of breast cancer's return," *Washington Post*, May 16, 2001, p. A10.

461 Ibid.

actually declined during the decades of DDT use.[462] They also showed that the occupationally exposed suffered no increase in the risk of liver cancer, thus refuting the environmentalists' claim that DDT caused liver cancer. Instead of this paper being widely read in the pages of a leading magazine, it was published in the relatively obscure journal *Ecotoxicology and Environmental Safety*. This bias of science reporting toward the sensational is also supported by relatively recent research that analyzed 1,812 randomly selected biological science papers from forty biology journals. Only nine percent of the papers contained "non-significant" results; the figure was even lower for the most prestigious journals.[463] But a non-significant finding in a study designed to confirm the results of a previous investigation is, in fact, highly significant.

So established are the beliefs about DDT that even today, the EPA does not denounce the falsehoods in *Silent Spring*. Instead, the Agency honors and enshrines those falsehoods. As stated on the EPA's website,

> *Silent Spring* played in the history of environmentalism roughly the same role that *Uncle Tom's Cabin* played in the abolitionist movement. In fact, EPA today may be said without exaggeration to be the extended shadow of Rachel Carson. The influence of her book has brought together over 14,000 scientists, lawyers, managers, and other employees across the country to fight the good fight for "environmental protection."

---

462 Deichmann, W.B. and MacDonald, W.E., "Organochlorine pesticides and liver cancer deaths in the United States, 1930–1972," *Ecotox Envir Safety*, vol. 1, no. 1 (1977): 89–110.

463 Csada, R.D., et al., "The 'file-drawer problem' of non-significant results: does it apply to biological research?" *Oikos*, vol. 76, no. 3 (1996): 591–93.

> Skeptics then and now have accused Carson of shallow science, but her literary genius carried all before it.[464]

Now, decades after *Silent Spring* was first published, the EPA tells us that the environmental movement swept into power through Carson's literary genius—not because of valid science or ethical reporting. Indeed, her literary genius did carry all before it. It helped rip away the health protections that DDT provided to well over a billion people. Over time, those inspired by Rachel Carson and Paul Ehrlich who acted against the public-health use of DDT ended up condemning millions to death and billions to illness by preventable diseases. The toll of this brand of environmentalism in loss of human life and human health has grown to enormous proportions, and yet, many environmentalists still will not acknowledge DDT's great service to humanity.[465]

Summary

Although DDT was first synthesized in a laboratory and is produced in factories, it is not an "unnatural" chemical. We know with certainty that many chemicals that are similar to DDT and that bioaccumulate do exist widely and naturally in the environment. We also know that no studies have been able to link DDT with harm to human health, even when people are

---

[464] Lewis, J., "The Birth of EPA," *EPA Journal*, Nov. 1985, http://www.epa.gov/history/topics/epa/15c.htm (accessed Apr. 14, 2009).

[465] "DDT turned from saint to sinner in 1962 with the publication of Rachel Carson's environmental manifesto, *Silent Spring*. The book appealed to the conscience of all nations that the environment was in grave danger and nature must be rescued. DDT became the symbol, the 'fall guy' for all the chemical pollutants infecting the environment." Desowitz, *Federal Bodysnatchers*.

exposed to it in large quantities. When used in small quantities in malaria control, DDT protects people from often deadly mosquitoes. The public-health benefits that DDT confers are considerable and far exceed any of the unproven, theoretical risks that people in the anti-DDT movement put forward.

Despite solid scientific evidence showing that DDT was not responsible for the destruction of wildlife, as we explained in chapter 4, environmental groups and those opposed to population growth were determined to stamp out any use of the chemical anywhere in the world. Environmentalists and those opposed to population growth willfully ignored scientific data that did not support their cause. As we see from their numerous writings and statements, some within the ZPG group were singularly misanthropic and sought to deny people a chemical that could save their lives.

Many people will naturally object to this characterization of misanthropy. However, it is undeniable that the nexus of environmental groups and those opposing population growth have done incalculable harm to public-health programs, and have done so in the face of good scientific evidence that contradicted their chosen position and in the knowledge that their work would harm disease-control programs. Indeed, some did so precisely *because* it would harm disease control programs.

The anti-DDT coalition managed to discard good scientific evidence and replace it with a political agenda, one that was and is on shaky theoretical and empirical ground. Through various legal processes in the United States, this coalition single-mindedly campaigned against DDT, and to the detriment of the world's poorest and most vulnerable people, it won. The EPA hearings into DDT were sound, balanced, and scientific, but were turned into a pointless farce when William Ruckelshaus overturned the presiding judge's ruling and banned DDT. The campaigns against DDT and its eventual banning would not have been possible without support from the general media and scientific journals. Bias in many publications, but most worryingly in *Science*, meant that the anti-DDT coalition could advance its cause, while those defending DDT were effectively silenced.

As we explain in the next chapter, the anti-DDT forces continued to campaign against DDT and other insecticides in public-health programs. They even turned the World Health Organization away from protecting people's lives from insect-borne diseases and toward their flawed and deadly agenda of environmentalism over public health.

Chapter 6

The Decentralization of Malaria Control

By the mid-1960s, rapid progress was being made against malaria. The number of people freed from the burdens of the disease was advancing toward one billion. Pilot studies conducted in many African countries revealed sharp declines, if not total interdiction, of malaria. Such studies held promise that disease rates could be cut, and in some areas, malaria could perhaps even be eradicated. Likewise, the implementation of eradication programs in twenty-one malaria-endemic countries of the Americas had greatly reduced malaria rates and practically eliminated malaria deaths. Similar results were being achieved in other regions of the world. For example, malaria had been eradicated in Taiwan and was being rapidly eliminated in Nepal.

Against this backdrop of improvements in the health and welfare of poor people in poor countries, few people could have foreseen how such spectacular programs would be extinguished abruptly—but they were. Decisions against malaria programs were being made at very high levels within governments and international bodies. At lower levels of international bureaucracies, deliberative bodies worked to develop reasons for those decisions. At even lower organizational levels, the scientists and bureaucrats, over a period of many years, had to remake the public images of those spectacularly successful programs. Basically, they began generating reasons for programmatic changes even though they had almost no role in laying out the agenda for those changes. To explain why and how those changes occurred, we will examine some of the important events, international policies, and funding trends of malaria eradication in the 1960s.

The Malthusian ideology of a world doomed by the growth of human populations reached critical mass during the 1960s. As discussed previously, Paul Ehrlich's book, *The Population Bomb*, was the tipping point for building public activism against population growth. However, influential international bodies were working toward control of population growth years before Ehrlich's 1968 publication. Sir Julian Huxley created the World Wildlife Fund in 1961 because of his concern that growth of human populations in Africa was endangering African wildlife. Malthusian ideology would have a devastating impact on the fate of the global malaria-eradication program.

UNICEF, USAID and Malaria-Eradication Funding

USAID and UNICEF were the two dominant organizations financing malaria-eradication programs. USAID provided funding, commodities, and technical help;[466] UNICEF mainly purchased commodities;[467] and WHO provided coordination and technical services. The seeds of change were planted in the ideologies and policies of these organizations in the 1960s.

The popularity of Malthusian ideology in the 1960s, led to increasing interest in the control of population growth in developed countries. As a consequence of this interest the United States initiated its own programs and support for international family planning activities during the 1960s. Similarly, concern about population pressures stimulated many developing countries to develop their own family planning

[466] Skerry, C.A., Moran, K., and Calavan, K.M., *Four decades of development: The history of U.S. assistance to Nepal 1951–1991* (Kathamandu, Nepal: USAID, 1991), 50–150.

[467] UNICEF, Notes on history of UNICEF "policy," UNICEF internal report for briefing of Labousse, who became Director of UNICEF in 1965.

programs. For example, fifteen of twenty-one malaria-endemic Central and South American countries started family-planning programs in the 1960s, compared to no countries starting programs in the 1950s and only two in the 1970s.[468] Popularity of the Malthusian ideology led to a belief within the UN that population growth was a "new menace" that was blighting "the prospects of social and economic development in the Third World."[469]

In 1961, President Kennedy signed the Foreign Assistance Act and created the U.S. Agency for International Development (USAID). Four years later, the U.S. foreign-assistance program began to support International Family Planning (IFP). According to the National Wildlife Federation, an environmental activist organization working to stop population growth, the United States became involved in IFP soon thereafter.[470] In the Foreign Assistance Act of 1967, Congress stipulated that funds would be provided for IFP in developing countries. Congress further earmarked $35 million for IFP in the FY 1968 budget. In 1968, Congress earmarked $50 million for IFP in the FY 1969 budget, and in 1969, President Nixon lauded IFP and urged UN

---

468 Barroso, C., *History of the family planning movement in Latin America and the Caribbean*, International Planned Parenthood Federation, available from: http://www.rhsupplies.org/news_events/membership_meetings/eighth_meeting.html, (accessed Sept. 25, 2009). Newman, *Malaria eradication and population growth*; Ehrlich, *The Population Bomb*.

469 Prestwich, W., *UNICEF: Guilty as charged*, http://www.lifesitenews.com/waronfamily/unicef/unicef.pdf, (accessed Sept. 25, 2009) Subquotes taken from: Black, M., *The Children and the Nations: The Story of Unicef* (Sydney, UNICEF: 1986), 241.

470 National Wildlife Federation, "The importance of international family planning assistance", http://www.nwf.org/globalwarming/pdfs/IFP.pdf, (accessed Sept. 25, 2009)

leadership on the issue.[471] $75 million was earmarked for IFP in 1970 and a further $100 million in 1971.[472]

By 1972, population assistance had become a line item in the foreign-aid budget, with a $125 million appropriation. A decade later, President Carter proposed $345 million for population control, but Reagan reduced the budget to $211 million. In the 1990s, however, the IFP budget increased to levels above $400 million, and funding presently hovers around half a billion dollars annually.

UNICEF became involved in family planning around the same time as the United States did. In 1965, UNICEF Executive Director Henry R. Labouisse submitted to UNICEF's executive board a proposal for a "possible role of UNICEF in family planning."[473] Labouisse's proposal instigated a fight that split the board. Some board members thought "it would be wrong for UNICEF to depart from its mandate of saving children to engage in activities to prevent them from being born."[474] Labouisse's proposal was withdrawn the same year, but UNICEF sought support from the UNICEF-WHO Joint Committee on Health Policy on the "best way in which UNICEF might participate in programs of family planning."[475] In 1967, the Joint Committee submitted a report to UNICEF's executive board seeking approval for UNICEF to participate in IFP. The report made the following arguments:[476]

---

[471] Gibbons, J.H. and Coen, A., *Forty years of U.S. international family planning*, Jan. 3, 2005, http://www.populationaction.org/Publications/Reports/40_Years_of_U.S._International_Family_Planning/40_Years.pdf, (accessed Sept. 25, 2009)

[472] Report of Controller General of the United States, April 21, 1972, available at http://archive.gao.gov/f0302/096590.pdf (accessed August 17, 2009)

[473] Prestwich, W., *UNICEF*

[474] Ibid.

[475] Ibid.

[476] Ibid.

- "Birth spacing was a health service for both mother and child."
- UNICEF's family-planning programs should be "an integral part of comprehensive health services for mothers and children."
- "The need for rapid expansion of health services to provide scope for family planning as an integral part of them should be recognized."

UNICEF's executive board accepted the report in June 1967,[477] giving UNICEF approval to participate in IFP.

The 1960s saw a rush to judgment about population growth in developing countries. A stampede of policy changes addressed these population concerns and converted them into rapid budget growth for population-control programs. Funding for population control quickly outpaced maximal historical levels of funding for disease control (e.g., peak USAID spending on malaria eradication was $85 million over a three-year period),[478] and UNICEF, with help from WHO, began to reconfigure its policies so that it could participate in the growing flow of funds for population control. UNICEF ended its malaria-eradication support by 1964, gradually phasing out its support of $4.1 to $8.8 million annually, and took up a leadership role in the movement to stop human population growth.[479]

So what happened? Why did UNICEF pull its support for a public-health program in the mid-1960s that specifically improved health and welfare for untold millions of children? Certainly, UNICEF could have begun work in IFP while continuing to support malaria-eradication programs. Just as beginning work on a water-treatment plant is not an excuse for

---

[477] Ibid.

[478] Siddiqi, J., *World health and world politics: The World Health Organization and the UN system* (Columbus, SC, The University of South Carolina Press, 1995), 172.

[479] Ibid.

stopping trash pickup, the beginning of IFP should not have precipitated stoppage of funding for malaria eradication. A 1965 internal UNICEF briefing document explains the reasons for changes in program support. The executive director of UNICEF, an American named Maurice Pate, died January 19, 1965.[480] The new executive director, also an American, was Henry Labouisse. The assistant executive director at that time, E. J. R. Heyward, wrote a briefing document for Labouisse on March 17, 1965,[481] describing how UNICEF bought supplies and equipment for programs which, in turn, received technical assistance from WHO. He described the failings of WHO budgetary cycles and elaborated on how WHO tried to get UNICEF to pay WHO for its technical experts that helped the national eradication programs. The briefing document shows how this high UNICEF bureaucrat was dissatisfied with UNICEF's role in WHO's malaria program:

> For some, UNICEF is the supply agency. But it is the specialized agency [in this case, WHO], and not UNICEF, that determines how many project personnel are needed; the agency will not give technical approval unless adequate international personnel [are] provided for. Moreover, this position has been abused by requests for excessive personnel for whom UNICEF would pay; views are more reasonable if the agency is paying. If there is to be real partnership in assisting the country, both agencies should contribute financially.[482]

---

[480] Maurice Pate Papers, 1904–1985 (bulk 1945–1965): Finding Aid, http://diglib.princeton.edu/ead/getEad?eadid=MC103&kw=, (accessed Sept. 25, 2009).

[481] Heyward, A.S.G., Notes on history of UNICEF: "Policy," Mar. 17, 1965, http://www.cf-hst.net/UNICEF-TEMP/Doc-Repository/docs/x0006rkf.pdf, (accessed Sept. 25, 2009).

[482] Ibid., 7.

He went on to explain, "Malaria is the outstanding example of WHO leadership. UNICEF became the major international financial partner in a worldwide eradication campaign launched in 1956 by WHO. This came to absorb as much as one-third of UNICEF's resources each year."[483]

Heyward described views expressed by UNICEF's board of directors as malaria eradication dragged on (in reality, eradication had been underway for only nine years when this was written) that UNICEF was investing in general health purposes rather than specifically in children and that "results of mass campaigns cannot be maintained without skeleton health services on the ground."[484] He opined, "Unfortunately, WHO's intellectual strength in the fields of public health, maternal and child health, and nutrition has never approached that of its resources in the fields of the various mass campaigns."[485] He also wrote about the philosophical differences between WHO and UNICEF programs:

> Malaria is a nice illustration of the logically different assistance principles of UNICEF and a specialized agency. For UNICEF it would be logical to assist malaria campaigns, where malaria is a major problem of children (in heavily infected areas, it is a major cause of child mortality), even if only control is attainable and not eradication, e.g. tropical Africa. The eradication objective of WHO requires malaria to be eliminated even from countries where it is a minor public health problem, e.g. Egypt and North Africa. The UNICEF Board laid down UNICEF's position clearly in January 1964: "UNICEF

[483] Ibid.

[484] Ibid., 8.

[485] Ibid.

may participate in new campaigns if . . . (ii) the country considers malaria a major problem in child health."[486]

Despite this assessment by UNICEF's executive board, UNICEF started to pull its support for malaria programs after 1964. Why? The answer to this question is found at the very end of Heyward's 1965 briefing document:

> *The major policy question now coming rapidly over the horizon is whether UNICEF will assist family planning directly.* This has been discussed in the Board, but not yet in the form of a proposal that UNICEF should give assistance, though a few countries have requested it.
>
> UNICEF has given considerable aid indirectly, this being "unconscious" and unrecognized as such—through aid to networks of health services, assistance to the education of women through community development, etc. These are very important forms of aid if one believes that the social and motivational prerequisites for family limitation are important, as well as the availability of simple means. Establishing the value of the individual child, and higher parental aspirations for children have historically been important factors limiting family size.
>
> Since this indirect assistance has not generally been recognized, we have the impression that *some European governments preferred to increase their contributions to "economic" development rather than to UNICEF,* because they felt that social progress was already in advance of its economic base, and that *the effect of UNICEF assistance was to increase the population explosion* (emphasis added).[487]

[486] Ibid.

[487] Ibid., 20–21.

Note the two main factors in UNICEF's withdrawal of malaria-eradication aid described in Heyward's concluding paragraphs. The new executive director acted promptly to get UNICEF in the money stream of the population-control movement and to eliminate its participation in the malaria-eradication campaign. It follows, from Heyward's concluding statement, that UNICEF's support of malaria programs was terminated to eliminate the criticism that UNICEF, by reducing malaria disease and death, was contributing to a population explosion in developing countries.

UNICEF was not the only organization that reduced or stopped supporting malaria programs. USAID was the recipient of rapid increases in budget for IFP. Almost concurrently, USAID also rapidly reduced its support for malaria programs. As described by Javed Siddiqi,

> USAID, which had over the past twenty years contributed $375 million to twenty-eight national programmes, changed its priorities and terminated its support for all MEPs (malaria eradication programs) except those of Nepal and Haiti. The decentralization of malaria assistance and the placing of its control in the hands of local directors of USAID missions was a harbinger for the precipitous decrease in MEP funding. Many USAID mission directors were more interested in endorsing family-planning programmes, which they felt would address the growing problem of over-population.
>
> By 1976 the international funds allocated to anti-malaria activities represented only 42 percent of those available ten years earlier; if the monetary depreciation over that period is taken into account, the 1976 funds amounted to less than 20 per cent of the former financial assistance.[488]

[488] Siddiqi, *World health and world politics*, 172.

The list of key malaria-eradication events of the 1960s also describes another theme that UNICEF and WHO would eventually impose on malaria-control programs: the need for a rapid expansion of health services to provide capacity for family planning. To meet this need, WHO and UNICEF would, throughout the 1970s, push the theme of bringing what were at one time well-developed malaria-control resources into the primary health-care system. Beyond this, both organizations began to push the theme of a community-participation approach to malaria control.

We can't be sure if the changes in malaria-control programs were motivated by a need to improve organizational capacities to meet the requirements of IFP, but for certain the initial attempts to move malaria control out of centrally controlled management structures and into community-based structures or general health services met with some opposition. A 1972 report to WHO's executive board states that "a member drew attention to the problems posed by the integration of national malaria eradication organization into the general health services." The report goes on to state that where a country had good prospects of success in eradication, "it was important to have an autonomous service with authority, a strong organization and powerful logistic means."[489]

In 1974, WHO's assistant director general, Dr. Bernard, cautioned against the abandonment of centrally managed control structures, warning that malaria control would be weakened if it were to be incorporated into basic health services where those services were insufficient.[490] By 1977, however, WHO had

---

[489] WHO, "49th WHO Executive Board Part II," official WHO records, 1972, no. 198, p. 22.

[490] Bernard states, "the integration of malaria control into the basic health services, logical as it seemed, had sometimes, where those services were insufficient, resulted in a less effective program, which lost its identity within the wider measures and received less financial support, with a consequent weakening of effect." WHO, "27th World

either forgotten or dismissed Bernard's arguments. In that year, WHO and UNICEF held a special discussion on the promotion of primary health care (PHC), which strongly advocated a greater role for PHC in disease prevention. The following year, a report by WHO's director general was unequivocal about the role of PHC in malaria control, stating that "there is no doubt that a countrywide malaria-control campaign cannot be carried out without the full participation of the public-health services and the promotion of self-help in the community. Therefore antimalaria activities must be built up as an integral part of the primary health care system, the front line of the health services."[491]

This is a remarkable statement considering the amazing successes of the international house-spraying programs. The director general's assessment does not reflect the historical record, nor is it supported by any data showing that malaria control could be effective within a primary health care system. Regardless, in May of that year, the World Health Assembly endorsed these views in its resolution on malaria control, stating that the World Health Assembly "urges Member States to reorient their antimalaria programs as an integral part of their national health program in accordance with the guidelines set out in the Director-General's report and to increase their commitments (fiscal, administrative and technical) against malaria, within their national development plans."[492] As stated before, we cannot be certain about the reasons for forcing changes in how malaria control was performed. However, we can be certain that there was no research or data that showed that malaria-control efforts would be improved by moving the

---

Health Assembly Part II," official WHO records, 1974, no. 218, p. 385–86.

[491] WHO, "Malaria control strategy. Report by the director general," WHO, 1978, A31/19, p. 4.

[492] WHO, "31st World Health Assembly," official WHO records, 1978, no. 248, p. 482.

activity into PHC or that it would be better to conduct malaria control through a community-participation approach.

It is perhaps useful to examine malaria-control events from the perspective of a WHO insider, Dr. Mohyeddin A. Farid, who was at one time the chief of epidemiology in the malaria division at WHO headquarters. Farid writes that the period from 1969 to 1980 was an era of de-eradication and anarchy.[493] He describes the beginnings of the dismantling of malaria-control programs and increasing burdens of malaria. The number of malaria cases reported to WHO in 1965, 1970, 1975, and 1980 for all countries, excluding China and sub-Saharan Africa, were 4,942,883; 7,009,156; 11,853,829; and 15,813,583;[494] respectively. Farid described certain events that, in his words, "hastened the de-eradication process and led to a rapidly deteriorating malaria situation."[495] Chief among these events was concern among economists that mass campaigns, like malaria eradication, were leading to a population explosion that would deter development and lead to social conflicts. According to Farid, this led industrialized nations to withdraw support for malaria campaigns while they increased support for family-planning campaigns.[496] The irony, as Farid saw it, was that disease-ridden societies would not comply with family-planning advice in the face of high child-mortality rates.[497]

Farid's paper also notes that the pressure against the use of insecticides, particularly DDT, in malaria control was promoted

---

[493] Farid, M.A., "Views and reflections on anti-malaria programmes in the world," *Kaohsiung J Med Sci*, vol. 7 (1991): 243–55.

[494] Ibid.

[495] Ibid., 248.

[496] Farid writes that industrialized nations "sensitized international bodies and industrialized nations against giving assistance to mass campaigns like malaria," and that they "supported heavily the family planning campaign." Ibid.

[497] Ibid.

in a "militant" way by the United Nations Environment Programme from 1972 onward.[498]

This brings us to Farid's next malaria-control era, the era of cover-up and resignation.[499] Farid describes the era from 1980 and beyond as a time when "the grumblings of malaria began to be heard in the speeches of some delegates to the WHAs."[500] In 1978, WHO issued the Declaration of Alma-Ata, which, among other things, declared that "an acceptable level of health for all the people of the world by the year 2000 can be attained through a fuller and better use of the world's resources." The Declaration also underscored the need for effective community action in setting priorities, making decisions, and planning and implementing strategies to achieve better health.[501] Farid stated that the Alma-Ata promoters were "heedless of the malaria problem after integrating the malaria services, and indifferent to its impact on the socioeconomic development which has been theoretically targeted by the Health For All Policy."[502]

Malaria scientists such as Farid began to question the theoretical underpinnings of the decision to integrate malaria control into basic, primary health care.[503] Yet, the Alma-Ata

---

[498] Farid writes that "birth of United Nations Environment Programme in 1970, promoted the militant drive against the use of insecticides, and through many conferences paved the way for legislations against the use of insecticides, particularly DDT (the cheapest and most effective) and recommended the expansion of the biological measures in controlling malaria vectors." Ibid., 249.

[499] Ibid.

[500] Ibid., 250.

[501] WHO, "Declaration of Alma-Ata," International Conference on Primary Health Care, Alma-Ata, USSR, Sept. 6–12, 1978, Geneva, http://www.who.int/hpr/NPH/docs/declaration_almaata.pdf, (accessed Sept. 25, 2009).

[502] Farid, "Anti-malaria programmes in the world," 250.

[503] Farid writes, "investigators began to think that Alma-Ata overtook the vertical malaria services without providing a proper mechanism to

promoters were clearly heedless of the harm caused, and any concerns raised were not sufficient to cause policy changes, because the 1985 World Health Assembly resolution 38.24, which called for malaria control to be developed as an integral part of national primary health care systems,[504] repeated and even strengthened the 1978 recommendation. Later that year, more objections to this policy were raised by delegates to a malaria conference in Papua New Guinea in 1986 when the changes were characterized as "attempting to put a square peg in a round hole."[505]

As revealed by the writings of the Papua New Guinea workshop, the malaria-control community was not behind the machinations of WHO and the WHA. Farid writes that the disquiet and "grumblings" of the malaria-control community had to be shut down by WHO as they tried to wish away any doubts about the Health for All by 2000 strategy.[506] In order to cover up these objections, Farid maintains that the scale of the malaria problem was trivialized and chemotherapy (case detection and treatment) was promoted in malaria control without any discussion of the dangers of that approach, which would include the development of drug resistance and, in some cases, poor

---

deal with the problem, and began to question on what basis the decision to integrate malaria programmes was made." Ibid.

[504] WHO, "38th World Health Assembly," WHA 38.24 Malaria Control, Geneva, 1985.

[505] WHO, "Intercountry Workshop on Control of Malaria as part of Primary Health Care," Madang, Papua New Guinea, Oct. 28–Nov. 1, 1985, Regional Office for the Western Pacific of WHO, Manila, Philippines, Mar. 1986, p. 1.

[506] Farid writes, "It was natural that with the worsening of malaria situation, and fearing the impact of the repercussions on engendering doubts about the infallibility of the Health for All by 2000 strategy, particularly after both Governments and WHO declared their pledge of full commitment, the grumblings of malaria had to be muffled. Certain tactics were made to cover-up the deteriorating malaria situation in the world." Farid, "Anti-malaria programmes in the world," 250.

tolerance of drugs.[507] Another tactic was to cover up malaria epidemics and use deceptive statistics so that by 1979, "the reporting was not representing the malaria situation, particularly when drug distribution and treatment became the principal weapons against malaria."[508]

By 2000, and unsurprisingly without any meaningful progress in achieving "health for all," WHO's Expert Committee Report noted that the decentralization of health systems, a process promoted aggressively by WHO since 1978, had been managed badly and had had a detrimental effect on malaria-control programs. The expert report doesn't accuse any particular party of being responsible for the negative consequences of decentralization, but the expert committee specifically notes that

> as a result of this decentralization, which often occurs without any input from malaria control program staff, programs that were previously highly effective have lost direction and focus. In addition experienced and dedicated personnel with specific technical expertise have retired or moved to other positions. Thus, programs have lost the capability to respond rapidly and effectively to changing conditions (e.g. epidemics) and their commitment and ability to guide operations at the community level.[509]

This assessment describes a global public-health disaster. Malaria experts knew that any malaria-control program run on a decentralized basis would be likely to simply deteriorate into a number of uncoordinated actions that would not amount to a public health program. In some African countries, health care

---

507 Ibid.

508 Ibid., 251.

509 WHO, "Expert committee on Malaria," WHO Technical Report Series 892, 20th Report, 2000, p. 13.

systems are decentralized and any central control is discouraged. For instance, democratization and decentralization are two primary policy objectives of the Ethiopian healthcare system.[510] WHO and UNICEF must take the largest share of responsibility for the badly managed decentralization of health care, as for decades this was a process that they tirelessly promoted.

A key rationale for PHC was, and is, to deliver a variety of health care services to the community. But poorly integrated programs that attempt to replace malaria control with PHC have ended up undermining malaria control without a functional replacement. In some cases, no doubt, bureaucrats sought to capture the resources of centralized malaria-control programs in order to strengthen PHC systems, which desperately needed improvement. Another potential goal may have been to strengthen PHC systems so that UNICEF could then use them to deliver family-planning services at the community level. Furthermore, some developing countries wanted to take more control of their own health care systems, and the programs driven by Western nations were seen as a form of health colonialism that was not appreciated by the elites of those countries. Nevertheless, developing-country leaders, along with the internecine turf battles of self-aggrandizing WHO bureaucrats and UNICEF's pursuit of the community-participation ideology, caused the demise of valuable malaria-control programs at the cost of health and welfare for hundreds of millions of people.

Before and after the 1985 move toward malaria control in PHC, there was significant growth in the ideology of community participation as the preferred approach to disease control. WHO emphasized methods of vector control that a decentralized control program could conduct. Basically, by taking

[510] Anon, "Health policy of the transitional government of Ethiopia, September 1993," Available from Earth Institute, Columbia University, http://cnhde.ei.columbia.edu/files/4/1/2004-Oct-Tue-040225.pdf (accessed April 14, 2009).

responsibility for conducting disease-control programs away from the government (by eliminating centralized control programs), WHO directed attention away from indoor residual spraying. International groups replaced those centralized programs with relatively ineffective control measures that individuals or, at most, communities could implement. Therefore, an entirely artificial framework for disease control came into existence that dictated kinds of control measures that could or could not be used. This new framework was not based on the science of malaria control.

In the push to promote PHC for malaria control, WHO, UNICEF, USAID, and others largely succeeded in eliminating many centralized malaria-control programs. As their success in eliminating these programs spiraled upward, so did malaria rates (see figure 3-8).

One might assume that WHO gradually recognized a need to find effective alternatives to the indoor residual spray programs that it had undermined. One might say that WHO and UNICEF had painted themselves into a corner because the ideology they trumpeted greatly constrained the search for acceptable alternatives to disease control. Thus, any new approach to disease control must be compatible with community participation and would need to fit within the integrated vector management (IVM) concept. WHO defines this concept as an approach that "takes into account the available health infrastructure and resources and integrates . . . chemical, biological [and/or] environmental" measures to "reduce or interrupt transmission of vector-borne diseases."[511]

---

[511] WHO Global Malaria Programme, "Integrated Vector Management," http://apps.who.int/malaria/integratedvectormanagement.html (accessed July 25, 2009).

Political Pressure against DDT in Malaria Control Mounts

While the anti-DDT campaigns were building in strength, support for DDT and indoor residual spraying (IRS) was waning within WHO and the WHA, and popular opinion was turning against DDT and insecticides in general. In 1969, as reports of insecticide resistance filtered in, the WHA began the process of scaling down the malaria eradication effort. Even though the concerns about insecticide resistance were misplaced given that repellency is the main mode of action for DDT, the WHA began a slow process of reexamining the global eradication strategy that would ultimately see it replaced with a strategy of malaria control. The WHA stated,

> complete eradication of malaria from the world remains a primary task of national public health organisations, and that even in the regions where eradication does not yet seem feasible, control of malaria with the means available should be encouraged and may be regarded as a necessary and valid step towards the ultimate goal of eradication.[512]

Even though the WHA assured that eradication was the ultimate goal, by 1978, Professor Raffaele Vannugli, one of Italy's representatives to the WHA, pointed out that "the word 'eradication' was tending to disappear from WHO documents and to be replaced by 'control.'"[513] But it wasn't purely in wording that the effort to fight malaria was changing. Vannugli went on to point out to the WHA that

---

[512] World Health Assembly, "22nd World Health Assembly," Boston, July 8–25, 1969, WHA 22.39.

[513] WHO, "21st World Health Assembly," Committee A meeting, official WHO records, 1978, no. 248. p. 484.

> malaria no longer occupied the place that it did some years ago; for example, in 1968 the total budget for antimalaria activities was US $6,544,316 [$36.7m 2005$] while in 1972 it was US $6,040,000 [$28.2m 2005$]. Subsequently antimalaria activities had been incorporated in the program of parasitic diseases, for which the total allocation in 1978 was US $8,233,845 [$24.6m 2005$]. Thus, even if the whole of the allocation were used for antimalaria measures it would represent a decrease in real terms, taking into account inflation.[514]

During the 1970s and '80s, opinions about the use of DDT seemed to differ widely within WHO. Some were based on science, while others indicated that the propaganda of the anti-DDT campaigners had done its work well. For instance, in 1970, a Mr. Wright,[515] from WHO's unit of vector biology and control, defended the use of DDT during WHO's executive board meeting. Wright pointed out that despite the concerns about DDT's effect on human health, "there had been no toxic effects recorded among the 200,000 spraymen employed and the 800 million population living in houses under spray over a long period of time."[516] In the same meeting, Canada's representative, Dr. Layton, informed the meeting that the Canadian prime minister, Pierre Trudeau, had issued a formal statement saying that "research showed the environment to be widely contaminated with DDT residues." Layton went on to argue that "it was only prudent, in view of the persistent use of DDT and the lack of clear cut evidence, to limit the

[514] Ibid.

[515] Most of the minutes and official records of WHO and the WHA did not record the first names of representatives and scientists.

[516] WHO, "Executive Board, 45th Session Summary Records," Geneva, Jan. 20–29, 1970, EB45/SR/9 Rev. 1, p. 155.

contamination of the environment."[517] The following year, Dr. Bernard, WHO's assistant director general, reminded the WHA that "indoor spraying of routine antimalaria operations does not involve a significant risk to man or to wildlife" and that "withdrawal of DDT from malaria programs would be fraught with great danger and is unjustified in the light of present knowledge."[518] Also in 1971, a WHO expert committee on malaria control defended the use of DDT, and WHO published a special report on DDT that was aggressively in favor of its use in malaria control.[519]

DDT also had its defenders among the member states that make up WHO. For instance, in 1972, Iran's representative to the WHA, Dr. Tabibzadeh, pointed out that DDT "is still the best available weapon until such time as a new and safe insecticide [can] be found."[520] During the 1975 World Health Assembly, Bangladesh's representative stated that his country was using DDT to great effect and had not had any problems with insecticide resistance.[521] Again, in 1980, Dr. Shwe Tin,

---

[517] WHO, "WHO Executive Board, 45th Session," Geneva, 1970, p. 154.

[518] WHO, "24th World Health Assembly, No. 194, Committee A, Malaria Meeting, Section 4.8 Vector biology and control," Geneva, May 4–20, 1971, p. 366.

[519] The 1971 WHO expert committee on malaria recommended that "every effort should be made to ensure the continued availability of residual insecticides, particularly DDT, for malaria eradication, which is an essential health activity." WHO 1971, "WHO Expert Committee on Malaria Technical Report Series No. 467," Geneva, p. 42. Furthermore, WHO published a special report on DDT by Wayland J. Hayes, Jr, that dismissed many of the claims made concerning the effect of the insecticide on human health. Hayes, W.J., Jr., "The degree of hazard to man of DDT as used against malaria."

[520] WHO, "25th World Health Assembly, Eleventh Meeting," May 22, 1972, p. 371.

[521] WHO, "28th World Health Assembly," official WHO records, no. 227, Geneva, May 13–30, 1975, p. 520.

Burma's representative to WHO's executive board, explained that "DDT was the only effective insecticide and the cheapest and it did not affect food in the house."[522]

As valuable as these interventions no doubt were, Iran, Bangladesh, and Burma were hardly influential or politically important compared with the block opposed to DDT. For instance, during WHO's 1969 executive board meeting, WHO noted with concern that the Scandinavian countries, Canada, and the United States were all taking action to "place severe restrictions on the use of DDT" and that this "had been a matter of concern to a number of developing countries."[523] In other words, WHO made it known to these developed countries that their actions would damage disease-control programs in poor countries.

As restrictions on DDT in wealthy countries increased its cost and decreased its availability for malaria control, poor countries became concerned. For instance, in 1971, the Indonesian representative to the WHA remarked that the price of DDT had increased by around $50 a ton over two years (an increase of around 10 percent).[524] A 1972 WHO report on proposed program and budget estimates said, "A member pointed out that DDT was still the insecticide used for eradication programs in most countries. However, certain countries had stopped manufacturing DDT and some Member States were experiencing difficulties in obtaining the required supplies."[525] The member asked what WHO's position was, to

---

522 WHO, "WHO Executive Board, 65th Session," Geneva, 1980, EB65/1980/REC/2, p. 225.

523 WHO, "WHO Executive Board, 47th Session, Part II, Appendix 14 'The place of DDT in operations against malaria and other vector borne diseases,'" Geneva, 1971, p. 176.

524 WHO, "24th World Health Assembly Part II," official WHO records, 1971, no. 194, p. 366.

525 WHO, "Executive Board, 49th Session, Part II," Official WHO Records, 1972, no. 198, Report on the Proposed Programme and Budget Estimates for 1973—Chapter II, p. 23.

which the director general replied that at a previous board session, he had issued a clear statement that "DDT presented no risk when used in malaria eradication programs, and that the discontinuation of its use would have grave detrimental effects on the program."[526] According to WHO records, this statement on DDT

> had the important effect of bringing about a worldwide acceptance of the vital necessity of continuing to use DDT for malaria eradication. As far as was known to the Organization, the major manufacturers of DDT were still producing and exporting this insecticide; lack of supplies was not impeding eradication programs. As was known, WHO had only limited resources for providing Member States with supplies of DDT, but a great deal of assistance was being given through bilateral aid. The Director-General expressed the hope that such bilateral assistance, which had proved of vital importance for many national programmes, would be maintained at the same level in the years to come.[527]

Unfortunately, the WHO director-general's hope for continued bilateral (country-to-country) assistance would be dashed. In 1976, the French representative announced at the WHA meeting that "a loss of faith in DDT had become apparent, perhaps because of its polluting properties and the emergence of resistance in certain anophelines."[528] In 1979, Bahrain's representative to the WHA, Dr. Khalfan, noted that "malaria was a serious problem, but that it should not lead to an excessive use of control measures, because intensive use of pesticides had been hazardous to the environment and [to] the workers

[526] Ibid.

[527] Ibid.

[528] WHO, "29th World Health Assembly," official WHO records, 1976, no. 234, p. 603.

applying those measures."[529] The pressure from developed (and by then malaria-free) countries certainly affected malarial countries. When discussing the use of insecticides, the WHO Malaria Action Program's 1980 report states that "in some countries where coordinated insecticidal spraying could be important in the control of malaria, there has been political pressure to restrain its use, although malaria itself could be seen as an environmental pollutant."[530]

As vector control was undermined, WHO advanced a policy for the immediate treatment of malaria cases as opposed to the preventive measures entailed in spraying houses with DDT. While addressing the 1983 WHO executive board's seventy-first session, the Malaria Action Program's director, Dr. Najera-Morrono, stated that malaria control was best achieved "mainly through the use of drugs for chemotherapy rather than vector control."[531] By 1997, after years of pressure against the use of DDT, mostly from developed nations and long-term WHO staff members like Najera-Morrono, the WHA passed Resolution 50.13, which calls on countries to "take steps to reduce reliance on insecticides for control of vector-borne diseases through promotion of integrated pest-management approaches in accordance with WHO guidelines, and through support for the development and adaptation of viable alternative methods of disease vector control."[532]

That the WHA produced such a resolution is hardly surprising when one considers the advocacy against DDT that year. In January 1997, WHO's executive board, which sets the

---

[529] WHO, "32nd World Health Assembly, Verbatim Records of Plenary Meeting," p. 134.

[530] WHO, "WHO Executive Board Review of the Malaria Action Program," 1980, EB67/PC/WP/7 appendix 2, p. 3.

[531] WHO, "WHO Executive Board 71st Session, Summary Records," 1983, WB71/1983/REC/2, p. 201.

[532] WHO, "50th World Health Assembly," 1997, WHA/50/1997/REC/1 Resolution WHA 50.13.

agenda for the WHA, invited two speakers to make anti-DDT presentations to the board, while not inviting anyone to defend DDT. One of the anti-DDT speakers was Marybeth Morsink of Consumers International, an influential advocacy group. According to the minutes of that meeting, Morsink stressed the link between DDT and human cancer and insisted that "DDT used in the tropics was transported through water and air to indigenous people living in the Arctic, putting them at risk." After calling for the precautionary principle[533] to be integrated into an agreement on persistent organic pollutants, she added her voice to a "call for reducing reliance on DDT for vector control" and to developing alternatives so as to "hasten the phasing out of DDT."[534]

During the 1997 WHA meeting on environmental matters, the United States, Canada, and the Netherlands all added their voices to the chorus of demands that DDT be phased out. Resolution 50.13 did not appear to go far enough for some WHA members, however. Dr. Nightingale of the United States questioned WHO about which countries were planning to incorporate "health into their sustainable development plans" as part of Agenda 21, the UN's sustainable-development plan, and rebuked WHO for not being sufficiently opposed to the use of DDT, saying, "Rather than promoting the continued use of DDT, WHO should be advocating its restricted use."[535] Canada's Dr.

---

[533] Goklany, I. M., *The Precautionary Principle. A critical appraisal of environmental risk assessment.* (Washington DC, Cato Institute, 2001) p. 2. A good definition of the precautionary principle, known as the Wingspread Declaration is as follows: When an activity raises threats of harm to human health or the environment, precautionary measures should be taken even if some cause and effect relationships are not established scientifically. In this context the proponent of the activity, rather than the public, should bear the burden of proof.

[534] WHO, "Provisional summary record of the sixteenth meeting," WHO Executive Board, 99th Session, 1997, EB99/SR/16, p. 17.

[535] WHO, "50th World Health Assembly summary records," May 1997, WHA50/197/REC/1, p. 1–2.

Jean argued for reduced DDT use on the basis that some DDT might leak out of the public-health field into other areas (such as agriculture) with "negative effects on the environment."[536] Ms. Stegeman from the Netherlands didn't even feel the need to justify her demand for the reduction in DDT use, simply saying that "all 12 persistent organic pollutants mentioned in the report, including DDT, should be phased out."[537]

If these moves within WHO and among its influential member states were not enough, the trend toward decentralized malaria control would make IRS and the use of DDT even more difficult. The move away from centralized malaria-control programs to decentralized ones has made any form of indoor residual spraying using DDT or any other insecticide much harder. While there are several highly convincing arguments in favor of decentralizing state-run services and giving maximal powers to local communities, when it comes to certain public-health programs, decentralization is counterproductive, reducing program efficiency.

Until 1966, India ran a highly successful malaria-control program using DDT. The transfer of malaria control to the general health services brought about a resurgence of malaria.[538] With this change, the program ran into problems of administration, finance, and a DDT shortage. Yet at a time when the IRS program in India was failing in part because of a lack of guidance and central control, the WHA adopted resolution 22.39 in 1970 calling for "the need to initiate malaria control schemes within the general health services."[539] This would result in even less control and a movement away from structures that were

---

536 Ibid., 2.

537 Ibid.

538 Pattanayak, S., et al., "Malaria paradigms in India and control strategies," *Indian J Malariol*, vol. 3, no. 4 (1994): 141–99.

539 WHO, "23rd World Health Assembly," Geneva, May 5–22, 1970, Resolution WHA 23.12.

working well in many countries, including several in southern Africa and almost all of the Americas.

Malaria control could have functioned within general health services provided the centrally directed structure of the malaria-control effort was preserved. However, that was not what WHO and UNICEF policy makers of that era envisioned. There was a firm determination to move malaria control into primary health care systems and to emphasize malaria control through community-participation programs. This determination was not based on any data or research that proved one could improve the success of malaria control through such changes.

Decentralized malaria control often masquerades as IVM, which sounds good in theory, but in reality has done little for disease control. WHO's support for IVM began at much the same time as the promotion of decentralized malaria control. The 1971 WHO executive board report on malaria control promotes the use of larvivorous fish, nematodes, predacious mosquito larvae, fungi, and bacteria as alternatives to insecticides.[540] In 1982, WHO promoted environmental measures, such as the regulation of water in streams and the elimination of stagnant water, before the use of any chemicals. The same report also promoted biological means of control, such as fish or biological agents that would kill mosquito larvae, ahead of insecticides for IRS, even though there was little evidence that biological control might actually work in disease control.[541]

Decentralized malaria control simply means that malaria control is conducted and managed at the local level, often with just one or two interventions. IVM, on the other, hand should involve in-depth vector control expertise, evidence-based decision-making and use of a whole range of disease-control

[540] WHO Executive Board, 47th Session, Part II, Appendix 14, "The place of DDT in operations against malaria and other vector borne diseases," Geneva, 1971, p. 180.

[541] WHO, 1982, p. 19.

methodologies. Resources almost never exist at a community level for an effective IVM program. It is therefore becoming common for endemic countries to characterize their malaria efforts as IVM programs, when in reality, they rely primarily on mosquito bed nets or, in some cases, just case treatment.

Flaws of Integrated Vector Management

Proponents of IVM question whether the future of malaria control lies in the use of insecticides or in alternative vector-control strategies.[542] To answer this question, we must remember that malaria is primarily a disease of rural people and of poverty, and therefore any approach to control must account for rural conditions.[543]

The IVM approach to malaria control basically relies on multiple methods of vector (insect) control. Superficially, this appears to be an intelligent and reasonable approach to disease control, but unfortunately, it is not, as IVM fails to discriminate between vector control and disease control. Vector control does not necessarily equate to a meaningful level of disease control, nor does disease control necessarily equate to a meaningful level of vector control. Vector control is a method specifically designed to eliminate or kill mosquitoes. Alternatively, disease control is designed to limit disease transmission by breaking vector–human contact and preventing biting. For example, any use of chemicals on house walls or chemicals on bed nets is a disease-control method. This is true whether the chemical functions by repelling or by killing insects. Likewise, any engineering method designed to reduce the entry of vectors into

[542] Roberts, D.R and Baird, J.K., "DDT is still needed for disease control" Pesticide Safety News, 2002, 5(4):1-5.

[543] There are, of course, exceptions, such as *Anopheles stephensi*, but the polluted waters in densely populated urban areas are generally unacceptable to malaria mosquitoes, as is the water in artificial containers.

houses is a disease-control method, as every vector insect that is prevented from entering or that is killed if it enters lowers the probability of efficient transmission of human disease.

Immature insects are not directly involved in the biological transmission of diseases like malaria, leishmaniasis, and filariasis. Thus, any method that targets immature insects is a method of insect control, not a direct method of disease control. This is true whether or not a chemical or a biological agent is used to kill the larvae. Methods that involve reducing mosquito populations by draining aquatic habitats, eliminating containers, or manipulating water levels in wetlands are also methods of insect control. Each method directly eliminates or kills insects—it does not exert direct control over disease. Given that we have little knowledge of the vector population densities needed to maintain transmission, it is almost impossible to know how successful insect-control methods will be in reducing disease. Transmission efficiency varies greatly among locations, environments, and vector species. It even varies according to the social and cultural characteristics of human populations. In areas where highly efficient mosquitoes maintain the disease, the level of mosquito population reduction needed to control disease can be very high.

The issue of scale in vector-control operations illustrates another flaw of IVM. Malaria is mostly a problem of poor rural areas, so any malaria-control method applied in rural areas must be affordable, easily applied, have long-lasting impact, and be applicable over large geographical areas. In contrast to urban environments, large unoccupied areas encircling single rural houses may harbor abundant clean water habitats for *Anopheles* larvae. This means that any method designed to kill or eliminate widely dispersed, rapidly reproducing, and highly mobile vectors will be prohibitively expensive. An example from southern Belize illustrates the significance of these relationships for malaria control.

Aquatic habitats within 500 meters (about 0.3 miles) of a rural house in southern Belize were mapped with global position recordings and ground observations. The data were then entered into a geographic information system to calculate the total surface area of larval habitats. As seen in figure 6-1, the total surface area within a circle of 500-meter radius is over 785,398 m^2. The three vector species present around the house used a variety of natural habitats for egg laying: *An. vestitipennis* used cattail marshes and swamp forests, *An. darlingi* used rivers and streams, and *An. albimanus* used sun-exposed, grassy pools. As it turned out, over 142,000 m^2 of aquatic habitat (surface area of water) were identified. Unfortunately, this example does not tell the complete story because malaria mosquitoes can fly much greater distances than 500 meters to arrive at a house. If we defined our area of concern around the house with a radius of 1500 meters, then we would need to apply control measures over more than 7 million square meters of terrain.[544]

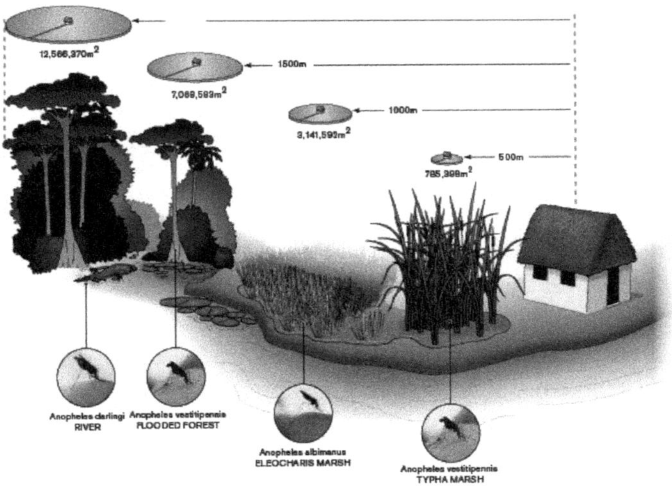

Figure 6-1: Surface area of water for mosquito propagation increases as a function of distance from human habitation.
Source: Roberts, D.R. and Baird, J.K., "DDT is still needed for disease control" *Pesticide Safety News*, 2002, 5(4):1-5

[544] Roberts, D.R and Baird, J.K., "DDT is still needed for disease control" *Pesticide Safety News*, 2002, 5(4):1-5

There exist multiple insect-control methods, including chemical insecticides, hormones, bacterial toxins, oils, and fish that are effective in killing immature mosquitoes. Water-management methods (drainage) might reduce mosquito production of the area's three vector mosquitoes in their different habitats. Water management, however, is seldom possible in rural areas because permanent drainage systems are costly and, in any event, there are often sound environmental reasons to oppose draining wetlands. Larvicidal treatments would also be problematic: they would have to be applied repeatedly throughout the year, as the three vectors are abundant in different seasons and some habitats dry and flood many times during the year. Furthermore, the cost of even a single treatment for a single household would be prohibitive.

Activism to stop public-health use of insecticides stems from environmentalists' desire to turn to more "environmentally friendly" means of control. Yet, effectively controlling malaria by manipulating aquatic habitats, as described for southern Belize, would cause a great deal of real environmental damage, as opposed to the fictional damage from insecticide sprayed on house walls. For example, exotic fish may prey upon or otherwise displace indigenous fish species and disrupt the environment in unpredictable and potentially far-reaching ways. The disastrous example of introducing cane toads in Australia for controlling beetles that threatened sugarcane crops should be a constant reminder to the advocates of larvivorous fish in malaria control.[545] Cane toads consumed many other beneficial insects, other frogs and even bird eggs and are now considered a serious environmental problem which the Australian government is attempting to deal with through multi-million dollar control programs.[546] Other means of eliminating the larvae, like

[545] Roberts, "DDT is still needed for disease control."

[546] Government of Australia, Department of the Environment, Water, Heritage and the Arts "The feral cane toad (Bufo marinus)," available from: http://www.environment.gov.au/biodiversity/invasive/publications/cane-toad/index.html (accessed August 17, 2009)

applying larvicides or surfactants across wetlands, would damage these sensitive and environmentally critical habitats. As Don Roberts explained in a 2002 article, "Attacking the larvae of malaria vectors in their wetland nurseries makes no sense from the perspective of either the environmentalist or the malariologist. Instead, the battle should be joined efficiently, effectively and safely where the infected mosquito lands upon human beings, in their dwellings."[547]

Under pressure from environmental activists and WHO, malaria-control specialists have reverted to a reexamination of many of the techniques that were marginally effective in the pre-insecticide days. Prior to the advent of synthetic insecticides, public-health officials had developed an array of vector-control methods. No single method or collection of methods dominated the pre-insecticide efforts because these methods were not very effective, as demonstrated by the historical data on disease rates up to the time that DDT became available. Unfortunately, the passage of time has not made those methods any more effective, nor has it changed the fact that most were designed for insect control, not disease control. Furthermore, the IVM methods trumpeted by WHO, the United Nations Environment Programme (UNEP), and environmental activists *are not* environmentally friendly. Almost any method of control applied over the landscape, as opposed to a control measure applied inside houses, will degrade natural environments.

Flaws of the Community-Participation Approach to Vector-Borne Disease Control

During the 1970s, the community-participation approach evolved as an ideology for moving the centralized malaria-control programs into primary health care (PHC) systems. As shown in WHO and WHA documents throughout the 1960s,

[547] Ibid.

1970s, and early 1980s, UNICEF and WHO in Geneva concocted this idea, and the WHA formally adopted it as a resolution in 1985.[548] But while centrally managed disease-control programs such as house spraying evolved through trial and error and through demonstration projects,[549] the community-participation approach to malaria control was not based on trial and error and was not founded on demonstrations of success.

The core component of centrally managed disease control was spraying DDT or other insecticides on internal walls at six-month intervals. DDT spraying was so successful that WHO's malaria-eradication program sent teams to conduct demonstration projects in several countries. These demonstrations provided firsthand evidence that house spraying actually worked and that a house-spray program could quickly reduce or eliminate malaria.

Case detection and treatment have long been a core part of malaria eradication. In the older, countrywide house-spray programs, patients sought medical care, cases were diagnosed and reported, complete antimalarial treatments were dispensed, and medical instructions were given for each laboratory-confirmed case. However, some countries, such as Mexico and countries in Central America, have now largely stopped spraying and instead place a strong emphasis on pharmacosuppression to control malaria.[550] The problem with this switch is that when drugs are distributed widely within the community as part of a pharmacosuppression program, patients stop seeking medical care and begin taking drugs on their own until they feel better, which increases chances for the development of drug resistance.

---

548 WHO, "37th World Health Assembly," Resolution and Decisions Annexes, Geneva, May 7–17, 1984.

549 Harrison, Mosquitoes, malaria and man, 234.

550 Pan American Health Organization, "Status of malaria in the Americas, 2004: A series of data tables," http://www.ops-oms.org/common/Display.asp?Lang=E&RecID=9466 (accessed Apr. 14, 2009).

Thus, historically, programs for mass drug distributions have accelerated the development of drug resistance.[551] Widespread drug distribution also means that cases are not diagnosed or reported and that governments no longer know the amount of disease within their communities.

An additional weakness of case treatment as a preventive measure is that symptoms can evolve slowly, so patients are often circulating parasites in their blood and infecting mosquitoes that come into their homes before symptoms of illness occur. This enables further transmission of malaria among the population. With treatment, the patients slowly become non-infectious to mosquitoes because parasites disappear from circulating blood. Yet they return to sleep in the very houses where they acquired malaria and where they served as donors of parasites to a new generation of mosquitoes. They will be sleeping in the same house days later when the new generation of mosquitoes returns to bite and infect them once again. It is for these reasons that spraying was the primary means of stopping disease transmission in almost all malaria endemic countries after WWII, as opposed to sole reliance on treatment of cases. Some countries, such as Guyana, had spectacularly successful malaria-control programs that initially had no element of treatment. In other words, a high level of control was achieved through DDT spraying alone.[552]

The Move to Insecticide-Treated Nets

With the constraints WHO and UNICEF had created, what evolved was an approach to malaria control applicable in a

[551] Verdrager, J., "Epidemiology of the emergence and spread of drug-resistant falciparum malaria in South-East Asia and Australasia," *J Trop Med Hyg*, vol. 86, no. 6 (1986): 277–89.

[552] Giglioli, G., "Changes in the pattern of mortality following the eradication of hyperdemic malaria from a highly susceptible community," *Bull Wrld Hlth Org*, vol. 46, no. 2 (1972): 181–202.

community participation program and compliant with IVM concepts. Thus they settled on insecticide-treated nets (ITNs) for people to sleep under during mosquito biting hours as their new-and-improved method of malaria control. The impact of joint UNICEF/WHO pressures on behalf of ITNs was revealed in a visit to Vietnam and Laos in 1992.[553] In discussions on September 25 in Vientiane, Laos, UNICEF's Anne Sutherland stated that it was not UNICEF policy to provide support for IRS or DDT, but pilot tests with ITNs were being supported in Laos, with UNICEF providing nets and WHO providing permethrin (a pyrethroid insecticide).[554] At that time, there were no stocks of DDT in Laos and no houses were being sprayed. The director of the Malaria Institute in Vientiane stated that the use of bed nets was effective, but the most effective malaria-control measure was DDT-sprayed house walls. UNICEF and WHO would provide no assistance to the government for acquiring DDT, however.

The visit to Laos was preceded by a visit to Vietnam. The personnel of the Institute of Malariology, Parasitology, and Entomology presented data on malaria rates and "demonstrated a corresponding increase in malaria rates as the numbers of houses sprayed with DDT declined during the last 4–5 years" before the 1992 visit. It was also observed during the visit that "the Vietnamese have . . . international support for purchasing permethrin and bed nets." In further discussion, they stated that they had been "unable to get international support for purchasing [DDT]."[555]

Years later, similar failures of international support were specifically described by government officials in Suriname and

[553] Roberts, D.R. "Trip Report. Viet Nam, Laos and Cambodia,"1992, 12 pages, unpublished.

[554] Ibid.

[555] Ibid.

Guyana,[556] and USAID used financial pressures to stop uses of DDT in Belize and Bolivia.[557] Today, many international groups deny opposing IRS and the use of DDT.[558] However, their exclusionary policies promoting ITNs and simultaneously pressuring countries to stop IRS programs and to not use DDT are a matter of record. These policies have been a major mode of operation for many years and continue even today.

For almost two decades, WHO, USAID, and some commercial interests invested heavily and invested only in ITNs in vain attempts to make them into more than they really are. Sadly, those very large investments did not grow out of spectacular demonstrations of success.[559] Rather, they resulted

---

556 Leslie Ramsammy, Minister of Health, Guyana, personal communication with Dr. D. Roberts, April 13, 2004.

557 Attaran et al., "Balancing risks on the backs of the poor;" U.S. Government Memorandum, "U.S. Litigation Document Search," from Mike Yates, Environment/Natural Resources, ARD to Jorge Calvo, ARD. Sept. 23, 1994.

558 For instance, the Roll Back Malaria Partnership's Global Malaria Action Plan (GMAP) now includes IRS as an important vector control tool whereas previously only the provision of ITNs was promoted and where IRS was discussed, it was usually couched in negative language, designed to undermine the intervention. Roll Back Malaria, "Global Malaria Action Plan" RBM, Geneva, 2008, available from: http://www.rollbackmalaria.org/gmap/

559 The problem with comparisons of cost-effectiveness between house spraying and use of treated nets is lack of solid quantitative studies. Most comparisons have been conducted with pyrethroids used for house spraying. Pyrethroids are expensive and such comparisons may show that ITNs are cheaper. However, DDT is cheaper than pyrethroids and when DDT is used, and the cost-effect almost always favors spraying over ITNs; Guyatt, H.L., et al., "A comparative cost analysis of insecticide-treated nets and indoor residual spraying in highland Kenya," *Health Policy Plan*, vol. 17, no. 2 (2002): 144–53; Conteh, L., et al., "The cost and cost-effectiveness of malaria vector control by residual insecticide house-spraying in southern Mozambique: a rural and urban analysis," *Trop Med Int Hlth*, vol. 9, no. 1 (2004): 125–32.

from imperatives that disease control be performed within decentralized control programs, that control measures be compatible with IVM, and that individuals or communities be able to carry out the control measures. Fortunately there seems to be some movement away from those policies, with more freedom for investments in infrastructure for organized malaria-control programs.

A considerable problem with ITNs is that they require user compliance; people must sleep under the nets for them to be effective. It is consistent across ITN evaluations that there is frequent failure of people to use the nets. Additionally, studies have shown that even when educational programs succeed in stimulating demand and use of ITNs, with time, use declines. In summation, user compliance is the Achilles' heel of total reliance on ITNs for malaria control. In contrast, once a house is sprayed in an IRS program, efficacy is not dependant on user compliance of the household.

Roll Back Malaria

The movement away from effective malaria control using IRS and DDT continued in 1996 when the WHA urged regional committees on malaria control to prepare regional and subregional plans of action. The same World Health Assembly urged members to "take action to participate fully in a reestablished action programme on malaria concentrating on ways and means of preventing and controlling the disease."[560] It was this resolution that paved the way for the 1998 formation of Roll Back Malaria (RBM), a broad partnership of various UN bodies and donor agencies to coordinate their efforts on malaria control. The founding partners of RBM were WHO, UNICEF, the World Bank, and the UN Development Program. Later on,

[560] WHO, "49th World Health Assembly," Summary Records and Reports of Committees, Geneva, May 20–25, 1996, WHA49/1996/REC/3, p. 78.

other organizations, such as USAID and the Italian Cooperation (the Italian aid agency), joined RBM. Under the leadership of Gro Harlem Brundtland, who had been appointed director general of WHO in 1998, RBM finally gave those UN bodies a vehicle to "coordinate"—that is, control—the malaria-control activities that successive World Health Assemblies of the 1970s had called for. RBM's primary goal was to halve the burden of malaria by 2010. Unfortunately, the early years of RBM were characterized by wasted time and missed opportunities, as the partnership had a narrow view of what should be done to control the disease.[561]

In the first six or seven years of RBM's existence, the partners relegated IRS to a second or perhaps third-line intervention for the prevention of malaria. First on their list of methods to prevent infection were ITNs and other insecticide-treated materials. RBM produced information sheets on malaria and malaria control and provided extensive information on ITNs, but almost nothing on IRS even though many countries, particularly those in southern Africa, use IRS as their main form of malaria prevention.

RBM's strategy to halve the burden of malaria by 2010 relies on six elements:[562]

1. effective management
2. rapid diagnosis and treatment
3. multiple prevention using insecticide-treated nets and other interventions
4. focused research
5. well-coordinated movement

---

[561] Attaran, A., "Where did it all go wrong," *Nature*, vol. 430 (Aug. 19, 2004): 932–33 and Snow, R., et al., "The global distribution of clinical episodes of *Plasmodium falciparum* malaria," *Nature*, vol. 434 (2005): 214–17.

[562] Roll Back Malaria Partnership, "Roll back malaria, country updates October 1998–June 2000," RBM, Geneva, 2000, WHO/CDS/RBM/2000.24, p. 6–7.

6. dynamic global partnership

RBM states that "a central principle governing the evolution of RBM at [the] country level is that there is no 'blue print,' implying that each country would chart its own, most suitable course."[563]

Yet, despite RBM's statement that there is no single preferred method of malaria control, it promoted ITNs above all other methods. More importantly, until recently, RBM's key funding partners, such as USAID, promoted ITNs almost to the exclusion of IRS. As described earlier, these policies have a long history and have effectively destroyed important disease-control programs around the world. It is a fundamental fact that countries that require funding for malaria have to align their policies with what the donor wishes to fund, even if that is not the best solution for malaria control.[564]

As we describe in the following chapter, following pressure from the U.S. Senate, in 2005, USAID began a process of changing its policy on IRS and the use of DDT.[565] However,

[563] Ibid., 3. This statement seems remarkably similar to one made in 1927 by the League of Nations Malaria Commission: "In every country and very largely in every area, there must be preliminary examination to ascertain what method is best suited to the local conditions. At present, it cannot be said that for malaria control there is a method of choice superior to all others." This same report went on to dismiss vector control in its entirety and to promote the use of quinine as the most effective method of malaria control. League of Nations Malaria Commission, 1927, p. 17.

[564] World Bank, Report No. PID 8820, Nov. 2000. For instance, a 2000 World Bank loan to Eritrea for malaria control was only granted on condition that the country reduced reliance on DDT.

[565] For instance, in Senate hearings before the Federal Financial Management Subcommittee in May 2005 and again in Jan. 2006, USAID was called on to support IRS. See http://hsgac.senate.gov/public/index.cfm?Fuseaction=Hearings.Detail&HearingID=abdc5968-19e5-44a1-a8a7-779ffc7b46f8 and http://hsgac.senate.gov/public/index.cfm?Fuseaction=Hearings.Detail&HearingID=a1170

prior to this, USAID's official policy toward DDT was that it did not exclude the insecticide from malaria control, but considered it an intervention of last resort, and then only subject to an environmental impact assessment (EIA). In correspondence with USAID, one of this book's authors (Tren) was informed that "in those (relatively rare) cases where DDT is truly needed for malaria control, the benefits of its use as a vector-control tool are considered to outweigh the risks the chemical presents to human health and the environment."[566] Thus, a donor could continue to argue, as USAID did, that the state of "last resort" has not yet arrived and therefore never fund DDT.[567] Furthermore, it is interesting that USAID requires an EIA even though DDT used for malaria control has never been shown to cause meaningful public-health or environmental harm. The requirement for an EIA demonstrates the success of the anti-insecticide campaigners in promoting their agenda. Given the minute quantities of DDT that malaria control safely uses and the vast numbers of people dying from malaria, a malaria-impact assessment would surely rule in favor of DDT. Yet, and contrary to good public-health practice, an EIA is still required before USAID can fund DDT use in a malaria-control program. Since USAID recommenced funding of IRS with DDT

cac-0d09-4470-9021-b691410e697b, (accessed April 12, 2009). In September 2005, the Senate Environment and Public Works Committee held a hearing on science and public policy and specifically addressed the issue of DDT and malaria control. See http://epw.senate.gov/hearing_statements.cfm?id=246814 (accessed Sept. 25, 2009).

566 Hirsch, Brian (USAID), June 20, 2001, email communication with Richard Tren .

567 Winston Churchill made an insightful comment that once held true for USAID policy. He noted that "Americans can always be counted upon to do the right thing, after they have exhausted all other possibilities." (QuoteDB - http://www.quotedb.com/quotes/2313). The tragedy is that millions of young children die while USAID is busy exhausting endless possibilities.

(as described in the next chapter), several such EIAs have been performed, all approving the use of DDT for malaria control. According to USAID sources, these EIAs are expensive and time-consuming and vastly delay the start-up of IRS programs. They do, however, provide valuable consulting opportunities for USAID contractors. The EIA is in fact a significant cost add-on for IRS programs that provides no benefit whatsoever to those who are suffering from malaria.

The formation of RBM coincided with a process led by UNEP to restrict or ban certain persistent organic pollutants (POPs), one of which is DDT. This process is discussed in greater detail in the next chapter, but it is important to note here that although WHO was a delegate to the various meetings and was fully informed of the process, it fell to the scientific community and some key countries that were using DDT to object to the proposed ban.[568] It is no coincidence that the objecting countries, including India and South Africa, are rich enough to fund their own malaria-control programs and choose to use DDT.

During this period in which the future of DDT for malaria control seemed most uncertain, WHO lent its voice to the many calls for the phaseout of the insecticide. The 2000 WHO Expert Committee on Malaria Report (report no. 892) recognized that IRS was used in many countries, but noted that some of these countries were reducing or phasing out IRS and were replacing DDT with next-generation insecticides, such as pyrethroids.[569] At the time, this was partially correct. Some countries, such as South Africa, were indeed evaluating the ongoing need for DDT and were attempting to comply with the 1997 World Health

[568] Two of this book's authors (Tren and Bate) attended several of the Intergovernmental Negotiating Committee meetings for the Stockholm Convention. WHO representatives never made an intervention that could be described as pro-DDT.

[569] WHO, "WHO expert committee on malaria, twentieth report," Report No. 892, Geneva, 1998, p. 37.

Assembly (50.13) resolution, which calls for countries to reduce reliance on insecticides for vector-borne diseases.[570] However as explained in Chapter 7, resistance to replacement insecticides and resurgent malaria meant that soon DDT would once again urgently be required.

A 1999 WHA discussion on RBM exposed how the demonization of DDT, along with the insistence that malaria control be community based, meant that IRS and particularly DDT had no place with WHO or with other international agencies.[571] Dr. A. Sanou Ira, a representative of WHO's executive board, stated that RBM's goal was to halve malaria cases by 2010, working "on the basis of agreed strategies; through community-based action against malaria . . . and by adopting appropriate measures against malaria."[572] This insistence that malaria control be conducted on the basis of agreed-upon strategies goes against the seemingly hollow assertion that malaria control is different in all areas and therefore requires different interventions. WHO had clearly chosen the "path of least resistance" with respect to malaria control, and that meant advocating ITNs only. At the same executive board meeting, the U.S. representative, Dr. Keusch, lauded WHO's efforts to reduce reliance on DDT and made the point that the "continuing controversy over the widespread use of DDT also meant that appropriate substitutes for vector control needed to be identified and evaluated."[573] Considering that this was a critical public-health issue, one would think that promoting reduced reliance on a particular disease-control measure would occur only after the appropriate substitutes were

---

570 WHO, "50th World Health Assembly, 1997, WHA 50.13 Promotion of chemical safety, with special attention to persistent organic pollutants," Geneva.

571 WHO, "52nd World Health Assembly," 2000, WHA52/ 1999/REC/3.

572 Ibid., 64.

573 Ibid., 65.

identified and made available—not after reducing reliance on the method.

At a subsequent RBM meeting, representatives from New Zealand, the Global Health Council, and the International Union for the Conservation of Nature (IUCN) all added their voices to the calls for a DDT phaseout. Speakers for the environmental groups were actually invited by the chairman to present their views, yet no supporter of continued use of DDT was invited to speak. The IUCN claimed that DDT was "a hazard not only to biodiversity but also to human health."[574] This claim went unchallenged even though, in fact, there is no evidence that DDT on house walls affects biodiversity or harms human health.

The 2000 WHO Expert Committee on Malaria may have correctly assessed the damage that decentralization had done to malaria control, but it missed the mark on IRS and DDT. The Expert Committee's main argument against IRS is that they claim it is financially unsustainable as a malaria-control intervention. It also argued against DDT due to unsubstantiated fears that it would harm the environment. Both arguments are false.

The Expert Committee's argument against DDT on the grounds of environmental contamination is a testament to the enduring success of Rachel Carson's campaign. The Committee admits that there is little or no evidence against the use of DDT on human-health grounds, but makes the nebulous, yet damaging statement that "It is clear, however, that in the broader context of sustainable development and in line with the principles agreed at the United Nations Conference on Environment and Development, WHO's concern now goes beyond human toxicology to the impact of DDT use on the

[574] New Zealand's Dr. Durham characterized DDT as "no longer appropriate, even in public health-mandated programs," and Ms. Newsome of the Global Health Council claimed that "DDT had been proved to have adverse effects on human health and well-being." Ibid., 67–74.

earth's resources and biodiversity."[575] The Committee would have been less disingenuous if it admitted that WHO's concern goes beyond its mandated function—the control of disease—to embracing almost everything that environmentalists want it to embrace. In what may seem an innocuous statement, WHO has seen fit to promote environmentalism and the unspecific concept of sustainability above protecting life and reducing disease. The consequences of this decision, which can only be understood if one knows and understands the history of malaria control, have proven to be both severe and deadly for people at risk from malaria.

As Dr. Amir Attaran, a well known malaria-control advocate, frequently explains, no one demands that garbage collection in his home town be sustainable. The residents simply decide that it makes sense for the local council to collect garbage, and they therefore pay for the services. Paying for services makes them sustainable. In contrast, WHO and key donors decided that IRS is unsustainable and therefore refused to fund it—making it, unsurprisingly, unsustainable. There is, in fact, no case to be made that IRS is unsustainable. Many countries, such as South Africa, Namibia and Swaziland have sustained such programs for decades, in spite of strong international pressures to comply with WHO guidance and dismantle their programs, and as a consequence have successfully controlled malaria. In any event, no public-health program, such as childhood immunizations, is inherently sustainable. All public-health programs require continuous funding, and the oft-repeated claim that IRS is not sustainable has no more merit than the claim that an ITN program is not sustainable. Both require continuous funding and support.

WHO, UNICEF, and various donor agencies' preferred strategy of ITNs is, according to these agencies, more sustainable. In 1998, at the formation of the RBM partnership,

---

[575] WHO Expert Committee on Malaria, WHO Technical Report Series 892, 1998, p. 39.

WHO estimated that there were between 300 and 500 million cases of malaria and between 1 and 2 million deaths every year. In 2009, some countries reported gains against malaria; however, after eleven years of RBM, the broad estimates of the number of cases and deaths are largely the same. This hardly amounts to a ringing endorsement of the strategies that RBM has pursued. Compare this performance with what was achieved in an equal number of years during the global eradication program, which saved millions of lives and freed almost a billion people from endemic disease.

The Expert Committee noted that IRS is the primary form of vector control "mainly in the Americas and [in] some countries in Asia,"[576] but neglected to mention that IRS is also the main form of vector control for almost every country in southern Africa. The report goes on to argue that improved malaria epidemiology and better surveillance should lead to "more selective vector control and further reduce reliance on indoor residual spraying."[577] This assessment properly classifies the overall emphasis of modern policies for disease control, whereby environmentalism is emphasized, not control of disease. Regardless, there is something to this argument. In the KwaZulu-Natal Province of South Africa, malaria-control program managers use geographic positioning systems to identify exactly where and when malaria cases occur. Site-specific data allow malaria-control programs to precisely target control measures at sites of actual disease transmission. However, while this allows them to spray less insecticide, it does not necessarily reduce reliance on IRS. Areas that use IRS need to be ever-more effective in order to protect neighbors and people in contiguous low-risk areas who are not protected by sprayed walls. Additionally, by targeting only the high-risk houses for spraying, the control measure must be very effective indeed to prevent the infections in high-risk houses from

[576] Ibid., 10.

[577] Ibid.

occurring and becoming a source of infection for nearby low-risk and unprotected households.

When discussing the prevention of malaria, WHO's Expert Committee report makes the following elaborate statement:

> The Global Malaria Control Strategy stresses the selective use of preventive measures wherever they can lead to sustainable results. Such measures should be aimed at halting the deterioration of the malaria situation, minimizing the wasteful use of resources and contributing appropriately to the development of health services, intersectoral cooperation and community participation.[578]

It goes on to say, "Selective vector control involves the targeted use of different vector control methods alone or in combination to prevent or reduce human-vector contact cost-effectively, while addressing sustainability."[579] What the Committee means to say is that it considers IRS to be unsustainable and would prefer that countries implement measures that do not require a degree of expert oversight and careful planning, such as ITNs and other insecticide-treated materials. The Committee states that

> non-selective coverage, as used for DDT and other insecticides in the past, is no longer a recommended strategy. Reduction of widespread coverage is still needed in the Americas, Asia and in some areas of Africa (where malaria transmission is focal and unstable, and in epidemic-prone areas). Given the financial and human resources involved, combined with the potential for vector resistance and environmental concerns, indoor residual spraying should be used only

[578] Ibid., 37.

[579] Ibid.

> in well defined, high or special risk situations. DDT is being phased out because of its previous widespread use in the environment, and the resulting political and economic pressure.[580]

In these assessments, the Expert Committee is indulging in pure nonsense. Reduced indoor residual spraying has irrefutably caused dramatic increases of disease in South America.[581] For example, the relationship between increasing disease and declining use of DDT was documented for the Iquitos area of Peru, and it was further exhibited by stunning reversions to malarious conditions in the Brazilian cities of Manaus and Belem with elimination of the DDT spray program.[582] One of the key recommendations of the report is that "large-scale operational use of insecticide-treated materials should be actively promoted, especially in areas of stable malaria transmission in Africa, south of the Sahara."[583] The report makes this confident recommendation despite a subsequent admission that "their [insecticide-treated materials] long-term effects on malaria morbidity and mortality still need to be investigated."[584]

---

580 Ibid., 38.

581 Roberts, D.R., et al., "DDT, global strategies, and a malaria control crisis in South America," *Emerg Infec Dis*, vol. 3, no. 3 (1997): 295–302.

582 Boletim Cientifico of the Fundação de Medicina Tropical do Amazonas. Ano X-No. 38. Abr/Jun-2004. Data presented on the high levels of malaria in the city of Manaus in 2003 through June 2004, with over 100,000 cases. When DDT was used for malaria control, neither the vector (*An. darlingi*) nor malaria cases were present in Manaus; Povoa, M.M., et al., "Malaria vectors, epidemiology, and the re-emergence of *Anopheles darlingi* in Belem, Para, Brazil," *J Med Entomol*, vol. 40, no. 4 (2003): 379–86.

583 WHO Expert Committee on Malaria, WHO Technical Report Series 892, 2000, p. 62.

584 Ibid.

In any event, there is good evidence that IRS, using DDT, worked exceedingly well in many countries in Africa. In fact, WHO's own data show that parasite prevalence fell from 79.5 percent to 1 percent among infants under the age of one in Western Sokoto, Nigeria, while DDT was used in malaria control during the 1950s. During the same period, parasite prevalence fell from 47 percent to 1 percent for the same age group in Tanzania (then known as Tanganyika). Where DDT was being used in Tanganyika, infant mortality fell from 165/1,000 in 1955 to 78/1,000 in 1957 and the overall death rate fell from 24/1,000 to 12/1,000 over the same period. Uganda also used DDT in a pilot project between 1959 and 1960 in the Kigezi district.[585] The spraying ensured that parasite prevalence for all ages fell from 22.4 percent to just 0.5 percent in those areas where DDT was sprayed three times a year and from 12.5 percent to 0 percent where it was sprayed twice a year.[586]

WHO's report is also wrong in that malaria programs generally do not use IRS non-selectively, but rather in targeted areas of active malaria transmission. Malaria-control programs do not generally have unlimited budgets, so they will use those budgets carefully to offer the best possible protection to the population. Financial constraints thus exclude non-selective coverage. The lack of financial resources directed to IRS is a direct consequence of donor agencies and the various UN bodies involved in malaria control choosing to direct their funds toward ITNs—an intervention for which the committee admits there is no long-term evidence of effectiveness—and toward almost any other intervention that isn't IRS. Finally, the political and economic pressure that the Expert Committee speaks of is largely a consequence of UN actions. It is exceedingly duplicitous that a committee appointed by WHO concludes that

---

[585] Kigezi district is now known as Kabale district in southern Uganda.

[586] WHO Executive Board, 42nd Session, Appendix 14, "The place of DDT in operations against malaria and other vector borne diseases," Geneva, 1971, p. 179.

a disease-control measure cannot be used because of financial and political pressures when WHO was one of the principals applying the financial and political pressures in the first place.

Summary

The 1960s witnessed an overwhelming growth in Malthusian ideology. New programs, new bureaucracies, and vast sums of new money flowed into international programs to stop population growth in developing countries by undermining malaria–eradication programs. From the mid-1960s onward, policies were changed to support the growth of primary health care systems and to provide for public-health needs through community-participation programs. A motivating force for these changes might have been to improve the capabilities of family-planning services at the community level.

The banning of DDT for agricultural use in wealthy developed countries and later in poorer developing nations had serious implications for malaria-control programs. Although many public-health experts within WHO and elsewhere defended the use of DDT and tried to retain its use in malaria control, the pressure from environmentalist groups and from the major donors to WHO to phase out DDT and other insecticides used in IRS programs was insurmountable. Despite the strong scientific evidence to support IRS and the continued use of DDT in malaria control, countries were strongly discouraged from continuing those programs. In fact, they were pressured to move away from centralized, dedicated malaria-control programs and to adopt a decentralized program structure that made running any kind of IRS program virtually impossible.

As once-powerful centralized malaria control programs were decentralized and dismantled, malaria rates grew. WHO and other leading UN and donor agencies responded to a growing malaria burden by forming the Roll Back Malaria partnership. Far from recognizing the folly of past policy decisions, the RBM partnership perpetuated bad policies and, for many years, advocated only insecticide-treated nets to control malaria while railing against IRS and any use of DDT. Failure was the inevitable result of such policies, and RBM currently grapples with reinventing itself and trying to become a relevant and effective partnership.

As we will explain further in the next chapter, those malaria-control programs that were financially independent from donor aid and had sufficient political backing to ignore WHO policies maintained IRS programs and continued to use DDT. South Africa's IRS program is an important example—not only because it reaffirmed the need for DDT, but because it gave some political cover to other programs in the region that sought to expand IRS and continue using DDT.

Chapter 7

The Seesaw of DDT Politics

Malaria control is politically charged—it has always been so. When Jesuit priests in Peru first discovered quinine in the early 1600s, many Protestants refused to take it, not because it wasn't effective, but because it was a Catholic discovery. The politics of malaria control intensified toward the end of the nineteenth century with the discovery of the malaria parasite and the discovery that mosquitoes transmit malaria. As soon as scientists understood what caused malaria and how it was transmitted, opinions differed on how best to control the disease. Some thought the only viable way of halting malaria was to treat patients, while others thought that eradicating the mosquito vectors of disease was the only sensible strategy.

Today, there is much wider recognition that both vector control and treatment are necessary to control malaria. However, today, the political divisions within the public-health community have more to do with the methods of control, and specifically with what role indoor residual spraying and DDT should play. Until recently, the donor agencies and UN agencies that provide money, advice, and even active malaria control were often deeply involved in these political games. To the detriment of malaria control, political considerations frequently ignore good science and the evidence of nearly six decades of what actually works.

During the 1990s, the World Health Organization continued to pass resolutions calling for the reduction of pesticide use in public health. As described in the previous chapter, discussions within the World Health Assembly on DDT continued, with growing pressure against its use in malaria control. Some countries, such as South Africa, began earnest attempts to comply with these resolutions and to phase out DDT. Even though South Africa was not dependent on donors for its IRS program, as a member state of WHO, it was required to reduce its reliance on DDT. In the mid-1990s, South Africa's Malaria Advisory Group (MAG), which formulates malaria-control policy, decided to replace DDT with synthetic pyrethroids. At the time, MAG based its decision on good science and the belief that it was doing the right thing.[587] Little did it know that insecticide resistance to the alternatives would develop soon afterward, and MAG would have to bring back DDT to control one of the worst malaria epidemics in the country's history.

While South Africa's strategy of reintroducing DDT dramatically cut the burden of malaria, WHO, Roll Back Malaria, and most other donors took several years to, sometimes reluctantly and with considerable prodding, recognize the role that IRS and specifically DDT could play in saving lives (information that had indeed been well documented for several decades). This chapter looks at some of the recent successes in malaria control, particularly in Africa, and describes how malaria scientists, government employees and activists in malaria control fought to keep DDT available and finally managed to change WHO and some donor policies toward the magical white powder.

DDT in South Africa

During the late nineteenth century and early twentieth century, South Africa considered malaria a significant health problem. Outbreaks of the disease were common in areas that are now considered completely malaria free, such as Durban. During most of the twentieth century, malaria was held at bay by

[587] Sharp, Dr. Brian, Medical Research Council., Apr. 2005, personal communication with Richard Tren.

a combination of good IRS programs using DDT, effective diagnosis and treatment, better standards of living, and improved levels of development. As people became wealthier and the country improved its health systems, people became able to seek treatment when infected and afford their own personal mosquito protection.[588] Collectively, these factors have restricted malaria to the subtropical regions of eastern South Africa that border the far-poorer countries of Mozambique and Zimbabwe.

However, in 1996, the South African government, acting on advice from MAG, decided to phase out DDT from its national malaria-control program. There were several factors behind this decision. Gradual changes to housing, e.g., painting of walls, within the rural malarial areas meant that malaria control needed some alternatives to DDT. DDT is less desirable in Western-style houses that have plastered and painted walls. The insecticide is far more effective on the traditional African houses constructed from mud and dung. DDT also leaves a white residue on the walls of houses, which, while useful for quality-control officers who can see at a glance whether a house has been sprayed properly, is often unpopular with homeowners.[589]

---

[588] This protection has rarely been bed nets, but doors and windows with mosquito screens and insecticide coils and sprays at night.

[589] In southern Africa, the IRS season begins in November and runs through March or April. South Africa has a well-established migratory labor system, a remnant of apartheid, whereby men, for the most part, leave their wives and children in rural areas to seek employment in the cities. These men normally return home for an extended break at Christmas, and their wives frequently used to plaster over the white marks left by DDT so as to keep the home looking neat and traditional, as their husbands would expect to find it. This poses a problem for malaria-control officers, as no sooner has a house been sprayed, than the DDT would be covered over and rendered ineffective. However, a recent study in India showed that DDT continued to be effective even after houses were replastered (Gunasekaran, K., et al., "DDT indoor residual spray, still an effective tool to control *Anopheles fluviatilis*-transmitted *Plasmodium falciparum* malaria in India," *Trop Med Int Health*, vol. 10, no. 2 [2004],: 160–68).

Also, while no South African government, either during apartheid or after the advent of democracy, can be said to be overly sensitive to international opinion, the ongoing international, and in some cases local, pressure against the use of DDT may have simply reinforced MAG's decision.[590]

What MAG had not anticipated was that the *Anopheles* mosquitoes would quickly become resistant to the synthetic pyrethroid insecticides that replaced DDT and that one of the most efficient vectors of malaria, *An. funestus*, would again transmit the disease. *An. funestus* feeds almost exclusively on man and is endophilic, resting with great frequency in human dwellings. This means that the probability that it will transmit malaria parasites from one human to another is very high. DDT spraying had kept *An. funestus* out of houses and prevented it from spreading malaria, but as soon DDT was removed, the mosquito once again fed on man and transmitted parasites.

As early as 1997, malaria cases and deaths started to rise. By 2000, in KwaZulu-Natal Province, traditionally the most malarial province, malaria cases had increased almost five-fold, from just over 8,500 cases to almost 42,000 cases.[591] Malaria deaths increased from 22 in 1996 to 320 in 2000. The number of cases completely overwhelmed health care facilities.[592] Along

---

590 In a conversation with a tourist operator, one of the authors (Richard Tren) was told that foreign tourists to the malarial areas, which also happen to be areas of outstanding natural beauty and the location of many nature reserves, were not comfortable with the idea that DDT was being used in a supposedly pristine environment. During the height of the malaria epidemic however, tourists operators soon learned that tourists care more about the risks from malaria than from DDT, as they stayed away from the game parks and rural areas.

591 South African Department of Health, "Malaria Updates, May 2000," Pretoria.

592 Ibid; When one of this book's authors (Richard Tren) visited some of the clinics in KwaZulu-Natal during the malaria epidemic in 2000, he witnessed scenes that one would expect to find in a war zone. There were not enough beds to cope with the number of cases. Those patients

with insecticide-resistant mosquitoes, the malaria-causing *P. falciparum* parasite developed resistance to the first-line drug treatment, sulphadoxine pyrimethamine (sold as Fansidar®). Drug resistance was in large part responsible for the high mortality rate, so in 2000, the Department of Health changed its common malaria treatment to Coartem®, an artemisinin-based combination therapy (ACT). ACTs are a highly effective treatment combining an ancient Chinese herbal remedy (artemisinin) with another modern drug (lumefantrine in the case of Coartem®) that has a different mode of action. The combination not only clears parasites effectively, but because the two drugs act differently against parasites, it delays the onset of drug resistance. Along with changing the drug therapy, the Department of Health also decided to reintroduce DDT to the IRS program because of the problem of mosquito resistance to the alternative insecticides.[593]

The combination of cheap, easy-to-use DDT and an expensive but highly effective malaria treatment ensured that within one year, the number of malaria cases fell by around 80 percent in KwaZulu-Natal Province.[594] Malaria deaths also plummeted as this one-two knock-out strategy took hold. The other malarial provinces, Mpumalanga and Limpopo, have since 2001 been able to keep malaria at bay by adopting the same strategies. Public-health experts widely acknowledge that South Africa's malaria-control strategy is one of the most successful in the world. In 2004, WHO even gave the South African

---

that were not sharing beds were forced to lie on the floor and suffer the agonizing fevers caused by malaria.

593 Barnes, K., et al., "Effect of Artemether-Lumefantrine policy and improved vector control on malaria burden in KwaZulu-Natal, South Africa," *PLoS Med*, vol. 2, no. 11 (Nov. 2005): e330.

594 Ibid.

Department of Health an award for the best malaria-control program in southern Africa.[595]

South Africa's long and successful history with DDT has influenced other countries in the region and provided political support for DDT and for IRS in general. The South African epidemic during the late 1990s and its demonstration of the continued need for DDT in large part secured an exemption for DDT under the United Nations Environment Programme's Stockholm Convention on Persistent Organic Pollutants (POPs). The tragic deaths of hundreds of people in South Africa during that epidemic is now helping to save thousands of lives elsewhere.

The Stockholm Convention

South Africa's malaria epidemic and the decision to reintroduce DDT came at a crucial time in the long political battle against DDT. At that time, the Stockholm Convention on Persistent Organic Pollutants was in the final stages of negotiations, with the last committee meeting scheduled for Johannesburg in December 2000. Malaria-control scientists were then engaged in a last-ditch effort to save DDT's use for malaria control.

The Stockholm Convention came out of a decision made in 1995 by the UNEP Governing Council (Decision 18/32, May 25) to develop a legally binding instrument to control certain chemicals. WHO supported the Convention in 1996 with a specific report on POPs and human health by WHO's director general. The Convention initially targeted twelve chemicals, known as the "dirty dozen," arguing that these chemicals "pose

[595] UN organizations such as WHO often have internal factions that differ on some key issues. For instance, the WHO AFRO office in Harare had long been in favor of IRS and increased DDT use. However, WHO's head office in Geneva largely pushed an anti-IRS and anti-DDT agenda.

major and increasing threats to human health and the environment."[596] One of these twelve chemicals was DDT.

The signatories to the Convention negotiated over five meetings held between June 1998 and December 2000. Initially, it seemed likely that the Convention would ban all uses of DDT, including public-health use. WHO's position on phasing out DDT lent support to UNEP and the major proponents of the Convention (mainly the wealthy, developed countries of the Northern hemisphere). The position of influential environmental groups, such as WWF and Greenpeace, was firmly behind banning DDT.[597] Although country delegates denied that they ever contemplated a total DDT ban when interviewed by the authors of this book in Geneva in September 1999 and in Johannesburg in 2000, it is hard to know what would have happened without the defenders of DDT. Political memories are often short, and it is highly likely that without pro-DDT actors, the Convention would have required a complete phaseout of DDT–a deadline of 2007 had been targeted by environmental groups including WWF, Greenpeace, and so on.[598]

Luckily, there was a small but dedicated group defending DDT, as countries that still rely on DDT were at a significant disadvantage during the Convention's negotiations. Most of these countries are less developed and do not have large budgets to send negotiating teams to Intergovernmental Negotiating Committee (INC) meetings. The INC meetings were

596 United Nations Environment Programme, "Decision 18/32 of the UNEP governing council: persistent organic pollutants," May 25, 1995, http://www.pops.int/documents/background/gcdecision/18_32/gc1832en.html (accessed Apr. 15, 2009).

597 In the face of significant challenges from malaria advocacy groups and developing country governments, these groups have subsequently modified their positions somewhat. They now claim they never opposed the public-health uses of DDT. The historical record and their own documents show otherwise.

598 World Wildlife Fund "Resolving the DDT Dilemma: Protecting Biodivestiy and Human Health", WWF, Washington DC, 1998

characterized by direct lobbying and vocal and aggressive campaigning of participants by environmentalist groups determined to rid the world of DDT and other POPs. Analysis of the attendance registers of the INCs shows the heavy representation of environmental NGOs, the G8 countries and the very light representation of malarial countries in Africa. On average at the seven INC meetings until the Convention came into force in 2004, 43 representatives attended from 32 sub-Saharan African nations and 3 individuals from 2 organziations defending DDT attended. This compares to an average of 72 individuals from the G8 countries (the United States, Canada, Britain, France, Germany, Japan, Italy and Russia) and 76 individuals from 43 different anti-DDT environmental groups.[599] As the chart below demonstrates, highly malarial sub-Saharan African countries were dwarfed by wealthy countries, which had already banned all the POPs in question, and by environmental NGOs determined to force through a DDT ban.

Invariably, malarial countries that use DDT could only afford to send one or two representatives, and these were not people from the departments of health that would understand the public-health implications of a DDT ban, but people from the government environmental agencies, or general trade attachés. In some cases, perhaps due to a lack of human capacity, the departments of health of countries that use DDT had not properly briefed the delegates, so they did not know that their

[599] In our analysis of environmental NGOs, we have included all those known to have taken a stance against use of DDT and to have been supportive of anti-DDT campaigns. We have not included NGOs in our analysis where we have been unable to find any anti-DDT or anti-insecticides publications or advocacy. We have not included chemical industry associations in the category of pro-DDT groups as it is not clear that these groups actually defended DDT, which by the late 1990s was a potential commercial threat to the sale of their newer products. As we describe in the text below, some in the chemical industry were open and up front about the fact they would approve of a complete DDT ban because it competed with their products.

countries still used the chemical to save lives. Unlike the delegations from most poor countries, the delegations from Europe and North America, were large, unified against DDT and could attend all the small committee meetings as well as the plenary sessions, making their case, often with the support and collusion of the many environmentalist NGOs.

Figure 7-1: Representation of different groups at the Stockholm Convention INC meetings

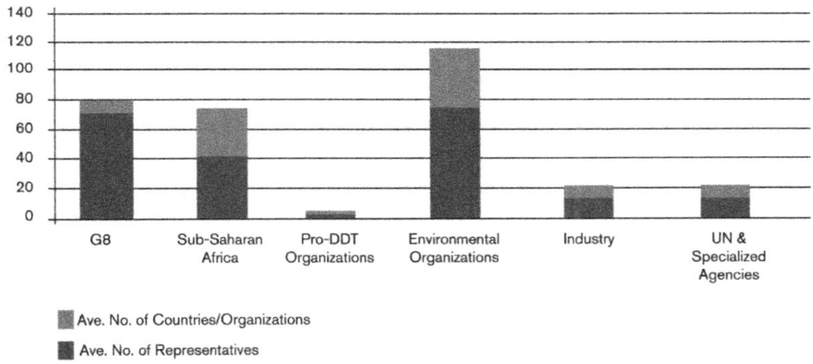

In spite of these problems, the efforts of some countries, particularly South Africa, and the scientists who rallied to DDT's defense secured an exemption for the use and production of DDT. The fact that the final INC meeting was held in South Africa meant that malaria specialists from the region and the South African Department of Health could lobby effectively and in numbers in favor of DDT.[600]

Prior to the Johannesburg meeting, lawyer and public-health expert Amir Attaran, Malaria Foundation International (MFI) founder Mary Galinski, and Don Roberts led a campaign to

[600] During the previous INC meetings, South Africa normally sent a delegation of two or three people. At the Johannesburg meeting, it could afford to send a larger team, including several public-health experts. Many malaria specialists also attended the meeting and forcefully defended DDT to the many country delegates and NGO observer groups.

protect DDT for malaria control. The Attaran/Galinski/Roberts petition gathered signatures from over 400 leading scientists, including several Nobel laureates. The letter was sent to all the delegates of the INC4 meeting in March 1999 and set the stage for a showdown between the malaria-control community and the environmentalists at the INC5 meeting.

The position taken by southern African governments, particularly South Africa, and the support from the scientific community allowed the pro-DDT lobby to secure an exemption for DDT. The Stockholm Convention listed DDT on its annex B, which allows any country that registers to seek exemption for either the production or use of DDT for disease control, instead of annex A, which would have required DDT's complete elimination. Although this outcome was clearly better for malarial countries than the original proposal to phase it out completely, using DDT is still not a simple matter for malarial countries. For instance, every three years, the Conference of Parties (the overall governing body), in consultation with WHO, is supposed to "evaluate the continued need for DDT for disease vector control on the basis of available scientific, technical, environmental and economic information."[601] While this evaluation may ensure that DDT remains on annex B, given the well-paid advocacy of environmental groups that work tirelessly to influence WHO officials and other signatories to the Convention, it is still possible that DDT could be phased out.

In addition to this sword of Damocles hanging over DDT, there are a number of stipulations within the Convention that make the ongoing use of DDT expensive and difficult for malarial countries. For instance, the Convention requires annex B countries to inform both the Secretariat (administrators of the Convention) and WHO on "the amount used, the conditions of

[601] Stockholm Convention on Persistent Organic Pollutants 2000, Annex B, Part II, paragraph 6, http://chm.pops.int/Portals/0/Repository/convention_text/UNEP-POPS-COP-CONVTEXT-FULL.English.PDF (accessed Apr. 15, 2009).

such use and its relevance to that Party's disease management strategy, in a format to be decided by the Conference of the parties in consultation with the World Health Organization."[602] These conditions may not seem terribly onerous for Western countries, but they could be burdensome for some poor countries, such as Eritrea or the Democratic Republic of Congo, that are struggling to rebuild their countries after civil wars and may have far-more-urgent national priorities than completing paperwork in UNspeak.[603]

The Convention also requires the parties to develop "regulatory and other mechanisms to ensure that DDT is restricted to disease vector control" and to "implement sustainable alternative products, methods and strategies" and "strengthen healthcare and reduce the incidence of disease."[604] It is, in fact, within the direct interest of those countries using DDT that regulatory mechanisms ensure that only public-health officers use it. If DDT procured for public health were to be used in agriculture, malaria control would be undermined by the diversion from its intended use. Indeed, most countries that use DDT already have specific legislation that restricts the use of DDT to public health.

Lastly, using the Stockholm Convention to compel countries to improve their health systems is, if nothing else, patronizing. Those countries that take the health of their citizens seriously will be striving to improve health care in any event and do not need an environmental treaty to compel them to do so. Furthermore, restricting the use of DDT in malarial countries by imposing bureaucratic hurdles to its use is precisely the sort of

---

602 Ibid., paragraph 4.

603 For instance, Eritrea has a life expectancy at birth of 52.7 years, only 46 percent of the population has access to proper sanitation and only 50 percent to essential medicines (UNDP Human Development Report 2003).

604 Stockholm Convention, 2000, paragraph 5(a).

measure that undermines good health care rather than improves it.

The Stockholm Convention came into force on May 17, 2004, after fifty countries ratified it.[605] Initially around thirty countries requested an exemption for use of DDT in malaria control, but as of mid-2009 and thanks to ongoing political and economic pressure, only seventeen countries remain on the list. Of these, only three countries—China, India, and Ethiopia—requested an exemption to both produce and use DDT.[606] The supposed benefits of the Convention have been notably absent. For example, the Convention encourages parties to the Convention to promote the research and development of safe alternative chemical and non-chemical products, methods, and strategies for countries using DDT.[607] In 2002, Dr. John Buccini, chair of the Intergovernmental Negotiating Committee of the Stockholm Convention, wrote,

> All Parties to the Convention are committed to developing alternatives to the pesticide [DDT]. The situation will be reviewed within one year of the Convention entering into force, and every three years thereafter, to see when DDT production and use may be eliminated. Considerable research and development efforts by all the world's countries will be needed if suitable alternatives are to be developed and implemented in a timely way.[608]

[605] For details on which countries have ratified the Convention, see www.pops.int. (accessed Sept. 25, 2009).

[606] Most of the DDT used in southern Africa is currently sourced from China, although India's Hindustan Insecticides has started to export DDT to certain African countries as well.

[607] Stockholm Convention, 2000.

[608] Buccini, J., "Briefing: Getting on top of the POPs," *Our Planet*, vol. 12, no. 4: 29, http://www.ourplanet.com/imgversn/124/buccini.html (accessed Apr. 15, 2009).

The Global Environment Facility (GEF), a global financing partnership established to address environmental problems, is the interim financial mechanism of the Stockholm Convention. Since its establishment in 1991, the GEF has provided grants of $8.6 billion and has attracted co-financing from partner organizations of $36.1 billion for various "environmental" programs and projects.[609] In 2006, thirty-two donor countries provided $3.13 billion to the GEF for projects between 2006 and 2010.[610] Despite this substantial funding, the GEF's investment into the search for alternatives to DDT has been disappointing. To date, it has only provided $37 million (0.5 percent of total grants provided) and secured a mere $47 million in co-financing from partner organizations (0.17 percent of total co-financing provided) "in support of DDT alternatives."[611] Closer examination of GEF-funded projects reveals that little or none of these funds were devoted to the search for chemical alternatives to DDT. Rather, GEF-funded projects have focused mainly on biological controls, environmental management, ITNs, and "combination[s] of the above with supportive insecticide use."[612] Thus it seems that GEF dispenses funds on the basis of an anything-but-insecticides policy, in violation of the Stockholm Convention itself.

Despite paltry public funding for alternatives to DDT and almost no allocation of these funds for chemical alternatives to DDT, the Conference of Parties (COP) to the Stockholm Convention and the Stockholm Convention Secretariat remain

609 Global Environment Facility, "GEF Funding.," http://www.gefweb.org/interior_right.aspx?id=50 (accessed Apr. 15, 2009).

610 Ibid.

611 Global Environment Facility, Stockholm Convention on Persistent Organic Pollutants, "Stakeholders' meeting to review the interim report for the establishment of a global partnership to develop alternatives to DDT: GEF support for the development of alternatives to DDT." Nov. 2008, http://www.pops.int/documents/ddt/presentations/GEF.pdf (accessed Apr. 15, 2009).

612 Ibid.

determined that all uses of DDT will be eliminated by 2020.[613] Yet, while UN environmental bureaucrats push ahead with plans to eliminate all uses of DDT, several African countries are, out of necessity and based on good evidence, restarting their use of DDT for malaria control.

African Countries Return to DDT

The South African epidemic, tragic though it was, not only ultimately strengthened the case for DDT and IRS in general and influenced the outcome of the Stockholm Convention, but also gave strength to those malaria-control programs in Africa that wanted to use DDT and expand their insecticide–spraying programs. During the late 1990s and early 2000s, the South African government launched several regional-development programs known as Spatial Development Initiatives (SDI). Each SDI has a different focus, depending on the comparative advantage of each region, the existing infrastructure, and the potential for development. The South African government launched a three-country SDI in the Lubombo region of Northern KwaZulu-Natal along with Swaziland and Mozambique. The focus of the Lubombo SDI (LSDI) is to improve agriculture and tourism by investing public funds in improving road linkages between the three countries.[614]

One of the limiting factors to tourism in the area, especially in the late 1990s, was malaria. In 1999, the LSDI began funding a joint malaria-control program in the three countries. The Regional Malaria Control Council (RMCC), which is made up of malaria program managers, scientists, and health specialists

[613] "Stakeholders' meeting to review the interim report for the establishment of a global partnership to develop alternatives to DDT." Nov. 2008, http://www.pops.int/documents/ddt/presentations/Secretariat%20of%20the%20Stockholm%20Convention.pdf (accessed Apr. 15, 2009).

[614] See www.malaria.org.za for a description of the LSDI.

from the three countries, manages the project. Its initial funding came mainly from both the public and private sectors in South Africa. The South African government provided R15 million (approximately U.S. $2.5M), and a private-sector development organization, The Business Trust, provided approximately U.S. $3 million (1999 dollars).[615] The Mozambican government also funded the malaria initiative to the tune of U.S. $580,000 (1999 dollars).

During the late 1990s, a UK-based mining group, Billiton Plc, began construction of a state-of-the-art aluminum smelter in southern Mozambique.[616] The initial investment in the smelter, known as Mozal, was over $1.2 billion (1990 dollars), with a further $800 million invested several years later.[617] Mozal is one of the largest single investments in Mozambique's history, and while the investors planned the construction in great detail, taking into account almost every eventuality, they underestimated the impact that malaria would have on the construction workforce. During the construction period, malaria cost Mozal over $2.5 million in terms of lost productivity and direct health costs.[618] In order to assist with malaria control, Mozal contributed approximately $580,000 to the LSDI malaria-

---

615 Business Trust, "Malaria: The regional malaria control programme," http://www.btrust.org.za/what-we-do/past-programs/impediments/malaria-the-regional-malaria-control-programme.html (accessed Apr. 15, 2009). The Business Trust is an initiative of 145 private companies in South Africa that works in partnership with the South African government to improve tourism and education.

616 Billiton Plc was originally known as Gencor, a South African mining company; however, it moved its primary stock market listing to the UK. In 2001, Gencor merged with the Australian mining company BHP and the new company became known as BHP Billiton.

617 BHP Billiton, "About Mozal," http://bhpbilliton.com/bb/ourBusinesses/aluminium/mozal/aboutMozal.jsp (accessed Apr. 15, 2009).

618 Tren, R. and Bate, R., *Malaria and the DDT story*, Institute of Economic Affairs, London, 2001, 83.

control program, which essentially funded an expansion of the control program to the Mozal plant.

While both South Africa and Swaziland had IRS programs that had been running for decades with well-trained staff and good health infrastructure, Mozambique was slowly rebuilding itself after a seventeen-year war that ended with democratic elections in 1992. Between the 1950s and late 1960s, a malaria-control program based on IRS using DDT operated in southern Mozambique and dramatically reduced the incidence of the disease.[619] However, by the time the LSDI program began in 1999, there had been little or no malaria control in Mozambique for almost thirty years, and the parasite prevalence in the south of the country reflected this lack of control. In Swaziland, thanks to the IRS program and rigorous efforts from the Swazi Department of Health to keep malaria cases low, the parasite prevalence at Mhlumeni, near the border with Mozambique, was just 2 percent in 2000. Over the border in Mozambique, the parasite prevalence was nearly 70 percent, and on Mozambique's southern border with South Africa, parasite prevalence among children between the ages of two and fifteen was almost 90 percent in 2000. At Makanis, on the South African side of the border with Mozambique, parasite prevalence in 1999 was 42 percent, falling to 25 percent in 2000 after the reintroduction of DDT.[620]

The LSDI bases its malaria-control program primarily on IRS, a logical choice given the historical and contemporary success of IRS in the region. South Africa and Swaziland use DDT. For the first few years of the LSDI malaria-control

---

[619] Mabaso, M.L.H., et al., "Historical review of malaria control in southern Africa with emphasis on the use of indoor residual house-spraying," *Trop Med Int Hlth*, vol. 9, no. 8 (Aug 2004): 846–56.

[620] Lubombo Spatial Development Initiative, "Malaria control annual report November 1999–March 2001," prepared for The Business Trust, Johannesburg, 2001, http://www.malaria.org.za/lsdi/Progress/annual_report_2001.html (accessed Apr. 15, 2009).

program, the Mozambican government refused to allow DDT to be used in its country, giving various reasons such as not wanting to offend donor agencies and concerns about environmental contamination. Instead of DDT, Mozambique used carbamates (a different class of insecticide that, unlike DDT, has a shorter residual life and shows no spatial repellent action), which, while effective, are considerably more expensive than DDT and also require twice-yearly spraying as compared with DDT's single spraying during a year.

As an aside, the historical use of DDT for malaria eradication resulted in perceptions that it should be sprayed on walls at six-month intervals in order to completely stop malaria transmission. The complete elimination of transmission was necessary to achieve the goal of eradication, but it was not necessary to achieve control. For this reason, there are inconsistencies in its modern-day use, with some countries spraying once per year for purposes of control and other countries spraying at six-month intervals in compliance with the old guidelines.

In July 2005, the Mozambican government changed its policy on DDT to allow its use in the country. It simply could not escape the outstanding successes in malaria control that resulted from DDT use in South Africa and Swaziland and clearly wanted its population to benefit as much as the citizens of neighboring countries were. In South Africa, malaria cases have fallen by 96 percent in KwaZulu-Natal Province and by 75 percent in Mpumalanga Province as of 2009.[621] By 2004, the average prevalence rate was below 0.25 percent and malaria incidence rates had fallen by more than 90 percent.[622] The results

[621] South African Department of Health, "National and provincial malaria figures," 2005, http://www.doh.gov.za/facts/index.html (accessed Apr. 15, 2009).

[622] Lubombo Spatial Development Initiative, "Progress in project to date in Maputo Province," http://www.malaria.org.za/lsdi/Progress/ProgressMaputo/progressMaputoProvince.html (accessed Apr. 15, 2009).

in Mozambique, even before the policy change on DDT, have also been very successful. Parasite prevalence in Zone 1 of the LSDI in Maputo Province fell from 62 percent at baseline to 1.9 percent in June 2007 and similar declines have been affected in other parts of southern Mozambique.[623]

The LSDI program shows that IRS can be a highly successful intervention in resource-poor areas that have little infrastructure, such as Mozambique. The strength of the LSDI program is that it is evidence based, meaning that it adopts interventions based on their proven efficacy. The LSDI also backs up its program with ongoing operational research to continually assess the efficacy of interventions and has concentrated on building up and strengthening local capacity to run the programs. So successful has LSDI malaria control been, in fact, that in 2003, the Global Fund to Fight AIDS, TB and Malaria provided a grant of $22.3 million to expand and strengthen the program. This funding was a great endorsement of the strategy and goes a long way toward ensuring the long-term sustainability of malaria control in the region.

Mozambique is not the only country in Africa that is moving back to IRS and DDT. For decades during colonial rule and for many years after independence, Zambia had an excellent malaria-control program that had reduced the disease to an almost insignificant problem. Zambia has one of the world's largest deposits of copper, and thanks to Zambia's reliance on its copper exports, the mines in the Copperbelt region in the north of the country had, from the 1930s, played an active role in malaria control by funding IRS activities that kept the disease at bay. In the early 1980s, however, that support dwindled along with the world price of copper and economic activity in the country. In the late 1990s, the Zambian government began the process of privatizing many of the state assets that were nationalized shortly after the country's independence in the 1960s. Decades of disastrous economic policies had left the

[623] Ibid.

economy almost in ruin and ensured that Zambia was one of the poorest nations on earth. Meanwhile, a combination of mismanagement and a crash in the world price of copper meant that many of the mines were no longer financially viable. The government linked the privatization process to other statewide, market-based economic reforms, but designed the process to breathe life into the mining sector. Luckily, the process also breathed life into the country's malaria-control program. In 2000, the newly privatized Konkola Copper Mine (KCM) restarted its malaria-control program and contracted Dr. Brian Sharp of South Africa's Medical Research Council to plan the malaria-control strategy. Sharp, who was instrumental in the success of the LSDI program and an advocate of IRS, based KCM's program on IRS using both DDT and synthetic pyrethroid insecticides. The malaria-control activities, which private sources entirely fund, cover approximately 31,500 dwellings, protecting about 350,000 people in the whole region, whether or not they are associated with the mine. In 2000, the program sprayed just under 80 percent of the dwellings with DDT and the remainder with pyrethroids.[624] Insecticide choice was based on type of dwelling. After the first round of spraying, there was a 50 percent reduction in malaria cases.[625] This success was matched with a further 50 percent reduction the following year. Malaria deaths in the mine clinics fell to zero in 2003, and the disease continues its decline in the region.[626]

The success of the KCM strategy has influenced malaria control in the rest of the country. Zambia's National Malaria Control Centre has restarted IRS programs in five other centers in the country. The funding for the programs initially came from the national government, using funds freed up by the World

---

[624] Sharp, B., et al., "Malaria control by residual insecticide spraying in Chingola and Chililabombwe, Copperbelt Province, Zambia," *Trop Med Int Health*, vol. 7, no. 9 (Sept. 2002): 732–36.

[625] Ibid.

[626] Data on file with authors.

Bank's Highly Indebted Poor Country scheme, and later came from a grant from The Global Fund to Fight AIDS, TB and Malaria. Initial results of the state-run IRS program are very promising and reflect a sharp decline in malaria cases.[627]

The other country in southern Africa to return to DDT in recent years is Zimbabwe. Like Zambia and South Africa, Zimbabwe had an excellent malaria-control program for many decades. However, the removal of DDT from the IRS program, along with all the other social and economic problems in Zimbabwe, led to a gradual increase in malaria cases during the 1990s. Starting in the late 1990s, the country went into a steep economic decline caused primarily by the Mugabe government's ruthless disregard for the rule of law and human rights. The disastrous economic and political policies of the Mugabe regime destroyed almost all formal employment in the country, drove millions of Zimbabweans out of the country, and by November 2008 produced an annual inflation rate of 89.7 sextillion (10^{21}) percent, making the currency utterly worthless.[628] The political and economic turmoil left the malaria-control program with few resources (although the Zimbabwean government has continued to fund its military and secret police generously). Partly as a result of the lack of funding, the country's once highly effective IRS program ground to a halt. In 2003, IRS coverage fell from 56.3 percent of targeted houses in malarial areas to just 3.4

---

627 Dr. Nawaa Sipilanyambe, then head of NMCC, Lusaka, Zambia, personal communication with Richard Tren, Nov. 25, 2004. The restarted IRS programs use DDT and concentrate in spraying houses in "hot spots" where the malaria-control program considers malaria transmission to be highest. The initial assessment of this strategy is that it has lowered malaria cases in the whole region, even though the spraying has been concentrated in the hot spots.

628 Hanke, S., "New hyperinflation index (HHIZ) puts Zimbabwe inflation at 89.7 sextillion percent," Washington, DC, Cato Institute, Nov. 14, 2008, www.cato.org/zimbabwe (accessed Apr. 15, 2009).

percent,[629] and consequently, malaria cases have risen sharply. Additionally, the general decline in the quality of medical care in the country means that effective treatment for malaria is often completely lacking.[630]

The other factor that contributed to the demise of Zimbabwe's IRS program is that during the 1990s, Zimbabwe removed DDT from its IRS program and replaced it with pyrethroids. According to statements made by then-Minister of Health Dr. Timothy Stamps, Zimbabwe's tobacco industry strongly supported abandoning DDT, as it feared that the international market would reject its product should any traces of DDT be found on it—an ironic concern since DDT is not a cause of human cancers, while tobacco's human carcinogenicity is well established. Replacing DDT with synthetic pyrethroids was a costly affair. Stamps explained, "The high cost of the alternative insecticide and the fact that more than one spraying program had to be carried out in epidemic areas for effective vector control had meant that, even with a 10-fold budget allocation, the government budget would only permit selective spraying."[631]

Partly out of self interest, from around 2004 South Africa sought to improve Zimbabwe's malaria–control program by donating DDT for IRS spraying in the southern region of Zimbabwe, which borders South Africa. Since mosquitoes don't recognize national boundaries, ensuring that malaria is under control on one side of the border greatly assists malaria control

---

629 Zimbabwean National Malaria Control Program, "Country Presentation," Southern Africa Malaria Control Joint Consultation and Planning Conference, Gaborone, Botswana, July 2004.

630 For more on the deterioration of the Zimbabwean health care system, see Africa Fighting Malaria's report, "Despotism and Disease," Mar. 2005, http://fightingmalaria.org/research.aspx?id=69 (accessed Apr. 15, 2009).

631 WHO, "52nd World Health Assembly." WHO, May 17–25 1999, (page 68).

on the other side.[632] In November 2004, the Zimbabwean government finally started to increase funding of malaria control and relaunched its IRS efforts in the other malarial regions of the country (mostly in the north) using DDT and synthetic pyrethroids. In February 2009, following the formation of a coalition government between the Mugabe regime and the opposition Movement for Democratic Change, USAID and the UK's Department for International Development (DFID) contributed emergency funds to undertake IRS in twenty high-risk districts with the aim of protecting over 400,000 households.[633]

On the back of South Africa's highly successful and controversial use of DDT, then-minister of health Manto Tshabalala-Msimang began advocating widely in favor of the insecticide. Her outspoken leadership on the issue in the media, at malaria conferences, and at public events gave impetus to other countries in the region to be equally forthright. One such country is Uganda. Approximately ninety-five percent of Uganda's population is at risk from malaria, and in 2006, the under-five mortality rate from malaria was measured at 137 per 1,000, or one in every seven children.[634] In 2003, Uganda's then-minister of health, Brigadier Jim Muhwezi, announced that his government would begin an IRS program using DDT as one of several interventions against malaria.

---

[632] Ever since the first democratic elections in South Africa, the ANC-led government has maintained friendly relations with Zimbabwe despite the ongoing human rights abuses in that country. The donation of DDT may also have been more about solidarity and friendliness with a government that assisted the ANC during the years of the anti-apartheid struggle.

[633] USAID, "USAID provides malaria assistance to Zimbabwe," Feb. 11, 2009, http://www.usaid.gov/press/releases/2009/pr090211.html (accessed Apr. 15, 2009).

[634] U.S. President's Malaria Initiative (PMI). "Country Profile, Uganda," May 2008, http://www.fightingmalaria.gov/countries/profiles /uganda_profile.pdf (accessed Apr. 15, 2009).

The choice of DDT is a good one and should have been non-controversial. The chemical is cheap and effective and had been used very successfully in pilot projects in Uganda in the past. However, to the detriment of malaria control, DDT use has been highly controversial and politically charged in Uganda. Local opposition politicians have made political capital out of the issue, portraying the minister of health as irresponsible and acting with little regard for the environment. One particular opposition politician, Ken Lukyamuzi, has accused the government of attempting to endanger peoples' lives by using DDT. At a public rally in Kampala, Uganda, in February 2007, Lukyamuzi stated that President Museveni was attempting to murder the people of Uganda by spraying DDT. Among his many fabrications and lies, Lukyamuzi told his audience that DDT would be sprayed on trees and in the environment and that it caused liver cancer, blindness, brain damage, and kidney failure. Lukyamuzi further misled the crowd by telling them that DDT was not mixed with water, but with benzene ethanol, chloroform, or petroleum. Perhaps most worryingly, this opposition politician incited people to violence, instructing them to "Get your saw! Your machete! Your axe! Greet the spraymen at your house."[635]

Possibly more damaging than the often-absurd statements made about DDT by disgruntled opposition politicians were the veiled threats from European Union officials in Uganda that the country's produce could be rejected from the EU should the government use DDT in malaria control.[636] Agricultural production for the export market in Uganda is unlike that of many other African countries in that it is mostly produced by small farms and produce is not kept in dedicated stores, but

---

[635] Excerpts of Lukyamuzi's speech can be found on YouTube.com, http://www.youtube.com/watch?v=4t2B-jbGIn4 (accessed Apr. 15, 2009).

[636] Masakeru, J., "EU warns on DDT," New Vision, Kampala, Uganda, Feb. 2, 2005.

frequently inside houses. This means that there could potentially be a close association between DDT and produce. Evidence is needed to determine the degree to which produce could be "contaminated." Nevertheless, the EU position has needlessly pitted agricultural interests against the public-health sector. The EU officials in Uganda ignore the fact that DDT can be managed successfully.[637]

In early 2008, with assistance from the U.S.-funded President's Malaria Initiative (PMI), the Ugandan government began IRS programs in the highly malarious districts of Apac and Oyam in northern Uganda. According to Stephen Mallinga, Uganda's Minister of Health, malaria cases and deaths had dropped by 20 percent in Apac and by 40 percent in Oyam shortly after the spraying.[638] However, soon after, agricultural exporters applied for a High Court injunction to halt the spraying on the grounds that it would harm exporters to the EU. On May 30, the High Court of Uganda ruled in favor of the agricultural exporters and halted the IRS program.

Environmental monitoring by the PMI found no traces of DDT in the soils and sediment of the DDT-targeted districts. Some DDT was, however, found on crop samples that had been stored inside homes sprayed with DDT. According to the IRS report prepared by PMI contractor RTI, the "concentrations of DDT ranged from 0.011 to 0.26 mg/kg (parts per million) in 55 percent of samples, with about 40 percent of the samples having

---

[637] This is confirmed by a letter sent by EU President Manuel Barosso to U.S. Senator Tom Coburn, dated Aug. 17, 2006, http://coburn.senate.gov/ffm/index.cfm?FuseAction=Files.View&FileStore_id=2398b4b7-e6de-4018-814e-b0f8a9f6b515 (accessed Apr. 15, 2009).

[638] Bugembe, A., "Health abandons DDT in north," *New Vision*, Aug. 5, 2008, http://newvision.co.ug/D/8/13/643070 (accessed Apr. 15, 2009).

levels above the European minimal residual level (MRL) of 0.05 mg/kg for agricultural products."[639]

As a result of the EU's policy, malaria control in Uganda has been disrupted and hampered. The Ugandan government is now forced to use alternative insecticides, against some of which mosquito resistance has been recorded. Although toxic resistance has also been recorded for DDT, as described earlier in this book, this resistance does not inhibit DDT's main mode of action, which is to repel mosquitoes, not kill them. Furthermore, even if agricultural produce were to be exported with the highest residues of 0.26 parts per million, this would in no way pose a human-health threat to the consumer. Imagine a consumer ingesting 11.25g of coffee (which would equate to two to three cups) with this maximum level of contamination. This would equal a dose of 0.0029mg of DDT—a level that is over 12,000 times below the "no observable effect level" as determined by long-term dosing experiments in humans.[640] These estimates are likely to be much higher than would be found in the real world because of DDT's exceedingly low solubility in water. This debacle is as clear a demonstration as any of the damage done to public health through the often-ruthless and wholly unscientific demonization of DDT.

In the late 1990s and early 2000s, the common thread among the most successful and far-reaching malaria-control programs in southern Africa has been that their funding was independent of WHO and/or donor agency politics. Either the national government, the private sector, or the Global Fund to Fight AIDS, TB and Malaria have funded the interventions in South

---

[639] RTI International, "Quarterly Report. Indoor residual spraying (IRS) for malaria control. Indefinite quantity contract (IQC) task order 1. October 1, 2008–Dec. 31, 2008." Feb. 2009, p. 48, http://www.fightingmalaria.gov/resources/reports/irs_fy09-q1.pdf (accessed Apr. 15, 2009).

[640] Hayes et al., "Evidence of safety of long-term high oral doses of DDT for man."

Africa, Swaziland, Mozambique, and Zambia. Those countries entirely reliant on UN or donor-agency funds for malaria control did not achieve similar successes against the disease.

Change in the Air and on the Ground

Over the decades, local and international politics has influenced malaria control greatly. The types of interventions sanctioned and supported by international agencies such as WHO, UNICEF, and the World Bank have not always been based on science and have been open to intense lobbying from outside of the malaria-control community. Despite the unequivocal success of DDT in malaria control in countries around the world, there has been a general reluctance, and in some cases an outright refusal, among the UN agencies and donor agencies involved in malaria control to support IRS and/or DDT. However, policy makers cannot escape hard facts and evidence forever. The years of solid data on the importance of IRS and the efficacy of DDT have finally resulted in a change in policy from WHO and from USAID, the largest and arguably most influential funder of malaria-control programs.

Persistent advocacy in favor of IRS and a rigorous defense of DDT began to pay off in Washington, D.C., when several influential congressmen took note not only of the tragic burden of malaria, but of the fact that a highly effective intervention was being systematically shut down. In a series of Senate and House hearings in 2004, 2005, and 2006, senators such as Sam Brownback and Tom Coburn and Congressman Ed Royce started asking hard questions about how U.S. taxpayers' money was being spent on malaria control and why IRS was not being used. One of this book's authors (Bate) exposed how USAID funding not only neglected IRS, but was opaque and largely favored U.S.-based consultants that consumed the majority of taxpayers' funds for malaria control. Less than 10 percent of USAID's budget for malaria control went toward buying actual commodities such as drugs, insecticide-treated nets, and insecticides that could save lives.[641] Senator James Inhofe chaired a hearing of the Environment and Public Works Committee to look at the science on DDT. The hearing, at which one of this book's authors (Roberts) testified, allowed U.S. lawmakers to better understand the implications of harm from basing decisions on politics and sentiment rather than on sound science. As the then-USAID Deputy Assistant Administrator for Global Health, Michael Miller, stated:

> Those criticisms spawned significant external pressure on USAID–especially from Congress–to focus on malaria and to focus on its problems. Although sometimes difficult to be in the crosshairs, within USAID we came to the conclusion that Bate, Tren, and Senator Tom Coburn were basically right, allowing us to argue internally for fundamental reforms. Malaria was a tough foe, but USAID's greatest problems in this case were institutional inertia, resistance to critical self-examination, questionable policies, and a lack of strategic prioritization.[642]

Miller continued:

> In the lead-up to the G-8 Summit in 2005, we were given an opening, and we produced a proposal that was for USAID truly unique, perhaps even revolutionary, with respect to its size and scope, its clearly articulated and measurable objectives, and its accountability to those who were to pay for it. It was not just the compelling idea of possibly saving the lives of millions

---

641 Bate, R. and Schwab, B., "The blind hydra—USAID policy fails to control malaria," Washington, DC, American Enterprise Institute, 2005, http://www.aei.org/publications/pubID.22355,filter.all/pub_detail.asp (accessed Apr. 15, 2009).

642 Michael Miller, personal communication with Roger Bate, September 03, 2009

of African children that had carried the day, but the fact that it would be able to demonstrate effectiveness–or be held accountable if it didn't.[643]

The quest from U.S. Congress for improved accountability and transparency of USAID as well as for funding the full range of anti-malaria interventions has led to some truly far-reaching reforms. In 2005, President George W. Bush announced a new malaria initiative, the President's Malaria Initiative (PMI). The PMI is USAID's lead malaria program and is designed as a $1.2 billion intervention over five years targeted at fifteen African countries where there was both an urgent need and an opportunity to have an impact.[644] From the outset, the PMI committed itself to funding IRS and even to purchasing DDT in those countries that sought to use it. To its great credit, the PMI is now supporting IRS programs in all fifteen countries and elsewhere, such as Zimbabwe. Since 2006, the PMI has sprayed many thousands of houses, protecting 24.7 million people.[645] USAID has now put IRS back on the menu and is attempting to undo the years of opposition against this intervention.[646] Until these reforms, USAID's position was that it would only promote DDT as a "measure of last resort," and as a consequence, it was never promoted at all.

The signs of a possible change in USAID policy surfaced in 2004. At the Southern Africa Malaria Control conference in Gaborone, Botswana, in July 2004, USAID contractor Gene Brantly announced that USAID was taking a more sympathetic

---

643 Ibid.

644 For a full description of the PMI, see http://www.fightingmalaria.gov (accessed Apr. 15, 2009).

645 President's Malaria Initiative, "PMI Results" available from www.fightingmalaria.gov/about/results.html. Accessed August 17, 2009.

646 An explanation of the PMI's position on IRS can be found at http://www.fightingmalaria.gov/technical/irs/index.html (accessed Apr. 15, 2009).

view of DDT by considering adopting the following position: "USAID supports the use of DDT for indoor residual spraying (IRS) in malaria control programs where alternatives either do not exist or are less effective and such use is in accordance with best professional practices."[647] However, it was not until more than two years later, in September 2006, that Admiral Tim Ziemer, the coordinator of the PMI, announced, "I anticipate that all 15 of the country programs of President Bush's $1.2 billion commitment to cut malaria deaths in half will include substantial indoor residual spraying activities, including many that will use DDT."[648] He went on to explain that "because it is relatively inexpensive and very effective, USAID supports the spraying of homes with insecticides as a part of a balanced, comprehensive malaria prevention and treatment program."[649]

Prior to the PMI and the reforms at USAID, U.S.-based contractors largely had control of funds for malaria control and had been less than transparent about how taxpayers' money was spent. Perhaps more important was the lack of any measure of impact of USAID funds on the incidence of malaria. Most contractors preferred to simply procure bed nets and use their own consultants or the private sector to distribute or sell them (or in some instances, just tell people that they should buy bed nets). By contrast, another malaria-control funding mechanism, the Global Fund to Fight AIDS, TB and Malaria, acts purely as a funding agency, providing funds for projects that an expert panel considers feasible and valuable. While there are problems associated with the Global Fund regarding technical review and the burdensome process of having funds released to purchase supplies, the Global Fund does not seek to advise a country on

---

[647] Brantley, Gene, EHP, Conference paper, Southern Africa Malaria Control Conference, Gaborone, Botswana, July 2004.

[648] WHO, "WHO gives indoor use of DDT clean bill of health for controlling malaria," Sept. 15, 2006, http://www.who.int/mediacentre/news/releases/2006/pr50/en/index.html (accessed Apr. 15, 2009).

[649] Ibid.

how to conduct its public-health programs; it simply provides funds for projects that an expert panel has vetted. Perhaps it is precisely because of this difference in structure that the Global Fund has, since its inception in 2002, provided funds for IRS programs and has allowed its resources to be used to purchase DDT.[650] Overwhelmingly, though, Global Fund support for malaria control has been directed to ITNs and not IRS. This is worrisome because there is a consistency in the dramatic malaria-control successes stemming from use of IRS, but not from ITN-only campaigns.[651]

The USAID malaria-control reforms went further than just a commitment to funding IRS and buying DDT for those countries that wish to use it. The PMI is vastly more transparent than any other USAID program or any other bilateral donor agency in explaining how, where, and by whom its malaria-control resources are being spent. In addition, the PMI has committed itself to measuring the effectiveness of its malaria spending not in terms of commodities procured or funds spent, but in terms of lives saved and cases of malaria averted. The reforms of the world's largest donor agency have been crucially important and should influence other donors and the private sector to similarly reform.

For a brief time, serious reforms were also put in place by WHO; it remains to be seen whether that level of leadership will continue. WHO is not a single, monolithic organization and

---

650 In two malaria-control programs, Zambia and the Lubombo Spatial Development Initiative, Global Fund money has purchased DDT.

651 For instance, to a large extent, the 2009 Global Fund report "Scaling up for impact, results report" downplays IRS. See http://www.theglobalfund.org/documents/publications/progressreports/ProgressReport2008_en.pdf (accessed Apr. 15, 2009). Furthermore, the Global Fund website overwhelmingly focuses on malaria treatment and the provision of ITNs, and mostly ignores IRS. Further discussion of global support for IRS can be found on Africa Fighting Malaria's website at http://www.fightingmalaria.org/article.aspx?id=1049 (accessed Apr. 15, 2009).

opinions within the group about IRS and DDT seem to differ widely. For instance, the Southern Africa Malaria Control (SAMC) program, based out of WHO's AFRO office in Harare, Zimbabwe, has for many years been pro-IRS and pro-DDT. Perhaps this is because the staff and officials of the SAMC are based near the real-world problems of malaria control. Health experts within multilateral (involving more than two parties) organizations that attempt to promote DDT have confided in this book's authors that they face opposition from some Geneva-based WHO officials.

Despite the obvious harm caused by anti-IRS and anti-DDT activism, WHO continued its marginalization of IRS and promotion of ITNs in its 2003 Africa Malaria Report. The report, which was jointly compiled by WHO and UNICEF, claims to be "based on a review of the best information available to WHO and UNICEF, from sample surveys and routine reports at the end of 2002."[652] At the time that the report was being compiled, South Africa had reintroduced DDT and strengthened its IRS program. Mozambique and Zambia had restarted IRS, and both Kenya and Uganda were debating the reintroduction of IRS. Yet, the report barely mentions IRS and focuses most of its malaria-prevention discussion toward ITNs.

Without a doubt, part of the reluctance to support IRS and/or DDT relates to the intense pressure and perceived political power of environmental activist organizations. Activist opposition is broadly published on the Internet, in the popular press and in UN documentation. UN alliance with activist groups was evident in 1999 WHA discussions, as described earlier. The committee chairman invited two activist organizations to make presentations while there were no invited presentations in support of DDT.

Reforms to WHO's malaria-control policy came in the latter half of 2005 when then-Director-General Dr. J. W. Lee appointed Dr. Arata Kochi to lead the Global Malaria Program

[652] WHO/UNICEF, "Africa malaria report 2003," 2003, p. 9.

(GMP). Kochi sought to distance the GMP from the Roll Back Malaria partnership, which was seen both internally and externally as a failed program. Although Kochi had worked within the UN and WHO system for many years, he had little experience in malaria control. This may have placed him in an ideal position to implement some much-needed reforms: without any history in malaria control, he had no need to justify any past positions that he may have taken.

One of Kochi's first tasks was to release new malaria-treatment guidelines in January 2006—the first time WHO had done this in twenty years.[653] Then on September 15, 2006, in Washington, D.C., Kochi launched WHO's new policy position on IRS. In his candid remarks, he explained how and why WHO had arrived at a position that strongly supported IRS and DDT. Kochi stated,

> I asked my staff; I asked malaria experts around the world: "Are we using every possible weapon to fight this disease?" It became apparent that we were not. One powerful weapon against malaria was not being deployed. In a battle to save the lives of nearly one million children every year—most of them in Africa—the world was reluctant to spray the inside of houses and huts with insecticides; especially with a highly effective insecticide known as *dichlorodiphenyltricloroethane* or "DDT."[654]

This statement marked an important change in the attitudes of those who set malaria policy in Geneva. Instead of being

[653] WHO, "Guidelines for the treatment of malaria," 2006, http://www.who.int/malaria/docs/TreatmentGuidelines2006.pdf (accessed Apr. 15, 2009).

[654] WHO, "Press statement by Dr Arata Kochi, Director of WHO's malaria department, Sept. 15, 2006, National Press Club, Washington, DC," http://malaria.who.int/docs/KochiIRSSpeech15Sep06.pdf (accessed Apr. 15, 2009).

inward-looking and insular, Kochi actually reached out to malaria experts and asked them what was needed. The importance of this change in policy and of having the imprimatur of WHO supporting IRS and DDT cannot be overstated.

However, Kochi's position did not go without criticism. Although Kochi specifically asked environmental groups to "help save African babies as you are helping to save the environment," his initiative had a mixed reception. Some organizations, such as the Sierra Club, disagreed with WHO's position, but went on to state that "Sierra Club does not oppose the indoor use of DDT to control malaria where it is critically needed and proven to be effective."[655] Other organizations, however, almost immediately rebuffed WHO. One group, Pesticide Action Network North America (PANNA), released a statement that included the following:[656]

> An issue that took center stage for NGOs was the controversy surrounding the World Health Organization (WHO)'s September 15th announcement giving DDT a "clean bill of health" for malaria control and promoting more widespread use of this old pesticide, directly undermining the goals of the POPs treaty. PAN International and allies (including IPEN and the International Society of Doctors for the Environment) staged a "Why DDT?" protest in Budapest and issued a joint statement of outrage calling for an investigation of WHO's controversial stance and highlighting our collective support for the POPs treaty approach to DDT.

---

655 Sierra Club, "Sierra Club's position on the World Heath Organization's promotion of indoor use of DDT to control malaria," Nov. 2006, http://www.sierraclub.org/toxics/ddt/ (accessed Apr. 15, 2009).

656 Pesticide Action Network North America, "From Budapest: Global experts ask 'Why DDT?'" 2006, http://www.panna.org/magazine/fall2006/newsWhyDdt.html (accessed Apr. 15, 2009).

> Now ratified in 132 countries, the POPs treaty calls for ultimate elimination of DDT but allows interim use in some countries with support for transition to safer and more effective alternatives.
>
> Many government officials at IFCS [the Intergovernmental Forum on Chemical Safety] voiced similar concerns about WHO's disturbing announcement and agreed to pressure the agency to clarify their policy on DDT. Since then, PAN has sent letters to individual WHO Board members expressing our dismay at the agency's stance, and promoting safer solutions for malaria control has become a focus for joint campaigns by PAN International and other partners led by PAN Africa.

This type of opposition is likely to persist. Over the decades, many people have ignored the science and turned a blind eye to the growing and unacceptable burden of malaria. There is no reason to suspect that this callous attitude will not continue. However, WHO's support for DDT and for IRS in general, along with the growing support from USAID and the Global Fund to Fight AIDS, TB and Malaria, means that 2006 was an exciting year for malaria control with great advances in policy and the implementation of those policies on the ground. In addition, private-sector support for IRS and DDT has grown over the years, and the position taken by WHO and USAID is likely to give the private sector further cover and authority to support such programs. Previously, some companies may well have shied away from a malaria intervention viewed as controversial.

Although policy changes have been vital, hard work lies ahead to implement IRS after decades of anti-IRS policies that stripped out expertise from WHO and other agencies. For IRS to work, WHO and other agencies involved in malaria control need sufficient funding and a firm, unwavering commitment from political leaders in both the donor nations and the malarial nations to support this life-saving intervention. Countries are using other insecticides in their expanded IRS programs, but they are not using DDT. Since the late-2005 turnaround at USAID and the September 2006 statements from WHO about the benefits of DDT, the only country to attempt to reintroduce DDT has been Uganda, and this program was halted.

The PMI should be lauded for its support of IRS and of DDT use in those countries, such as Zambia and Mozambique, that were already using it. But the lack of progress in supporting more countries in a roll-out of IRS using DDT is not the fault of senior management at the PMI. According to field sources and personal experiences, many USAID staff and contractors are still upset about the PMI's scientifically valid but nevertheless politically charged 2005 decision to use DDT once again. Some contractors undoubtedly saw the support for IRS and DDT as a threat to their budgets for ITN distribution, sales and education services, and ITN marketing. Another challenge is that the voices arguing against DDT have become more strident recently. Commercial entities involved in the marketing of alternative insecticides also contribute to the anti-DDT campaign. For example, a senior executive within Bayer Crop Sciences discouraged the use of DDT in Uganda, citing a possible threat to the country's supply of food for export while revealing that "DDT use is for us a commercial threat."[657] The environmental movement also continues to exaggerate the dangers of DDT.

Bias in the academic literature is another ongoing problem. A 2007 article in *The Lancet Infectious Diseases* in regard to the use of DDT alleges that superior methods for malaria control exist—without providing a single reference for this claim.[658]

---

[657] Beattie, A., "Commercial motive hinted at in restrictions on DDT," *Financial Times*, Sept. 29, 2005, http://www.ft.com/cms/s/0/0566ec34-3085-11da-ba9f-00000e2511c8.html?nclick_check=1 (accessed Apr. 15, 2009).

[658] Overgaard, H.J. and Angstreich, M.G., "WHO promotes DDT?" *Lancet Infect Dis*, vol. 7, no. 10 (Oct. 2007): 632–33.

These authors assert that DDT represents a public-health hazard by citing two studies that, according to a 1995 WHO technical report, do not provide "convincing evidence of adverse effects of DDT exposure as a result of indoor residual spraying."[659] Furthermore, the authors misrepresent those defending the use of DDT. They claim that supporters view DDT as a "panacea"—dogmatically promoting it at every opportunity—yet they do not provide any evidence to back up their opinion. Ad-hominem attacks on individuals defending DDT for malaria control, including the authors of this book, is a common tactic of groups within the anti-insecticide movement, as seen on Pesticide Action Network's DDT campaign website.[660]

UNEP is once again ramping up opposition to the use of DDT. At its third session, ending on May 4, 2007, "the Conference of the Parties of the Stockholm Convention requested its secretariat in collaboration with the World Health Organization and interested parties to develop a business plan for promoting a global partnership to develop and deploy alternatives to DDT for disease vector control."[661] The wording of this decision is important in that it only requests the secretariat to develop a business plan for *promoting* a global partnership, rather than requesting the secretariat to find the resources to develop the alternatives themselves. This could simply amount to the Stockholm Convention Secretariat spending taxpayers' money in a public relations exercise promoting a partnership. To date, the level of financial

---

[659] WHO, "Vector control for malaria and other mosquito-borne diseases," WHO Technical Report Series 857, Geneva, 1995, annex 1, p. 76.

[660] Pesticide Action Network North America, "Who's promoting DDT," http://www.panna.org/ddt/promoting (accessed Apr. 15, 2009).

[661] United Nations Environment Programme, "Report of the conference of the parties of the Stockholm Convention on persistent organic pollutants on the work of its third meeting," May 4, 2007, SC-3/2: DDT, http://chm.pops.int/Portals/0/Repository/COP3/UNEP-POPS-COP.3-30.English.PDF (accessed Apr. 15, 2009).

resources, both public and private, devoted to finding a true alternative have been minimal at best. Furthermore, UNEP and its partners have only focused on toxic modes of action and do not recognize that spatial repellency is the principal mode of DDT action.

Worryingly, Kochi, WHO's champion for IRS and DDT, has been sidelined. His outspoken and brusque manner won him few friends, as he himself noted in a *New York Times* interview: "The malaria[-control] community hates me . . .I said, basically, 'You are stupid.' Their science is very weak. The community is small and inward-looking and fighting each other."[662] The sudden death in 2006 of Kochi's most important champion, WHO Director General Lee, so soon after Kochi began reforming malaria control eventually undermined Kochi's tenure as head of the WHO's Global Malaria Program. For reasons unrelated to his professional work, Kochi was suspended. Regardless of the reasons for Kochi's removal, WHO has since been lacking in malaria-control leadership and the danger remains that any progress achieved in restoring IRS programs and in using DDT to save lives will be lost.

The rapid evaporation of any real leadership in malaria policies was visible in May 2009 after the publication of the Pine River Statement in *Environment Health Perspectives* (discussed in chapter 5). Shortly after the publication of this statement, WHO and UNEP issued a joint press release announcing that in line with a "goal of reducing reliance on DDT in disease vector control,"[663] WHO would work with UNEP and the GEF to combat malaria using "sustainable" (a

---

[662] McNeil, D.G., Jr., "An iron fist joins the malaria wars," *New York Times*, June 27, 2006, http://www.nytimes.com/2006/06/27/health/27prof.html (accessed Apr. 15, 2009).

[663] UNEP press release, "A Big Step Forward towards a DDT-Free World," May 6, 2009, http://www.unep.org/Documents.Multilingual/Default.asp?DocumentID=585&ArticleID=6148&l=en, (accessed Sept. 25, 2009).

word for which no definition is given) alternatives to DDT. What are these supposedly sustainable methods of malaria control? According to the press release, they include "securing homes with mesh screens [and] deploying mosquito-repellent trees and fish that eat mosquito larvae."[664] These same methods of control were discussed and promoted by WHO's Expert Committee on Malaria in 1979, which stated, "WHO is closely collaborating with the United Nations Environment Programme in carrying out investigations on environmental methods of malaria control and has, jointly with the Food and Agriculture Organization (another UN organization), reviewed proposals for integrated control of insects of both agricultural and public-health importance for consideration by UNEP."[665] After the publication of the 1979 report, malaria cases rose around the world, claiming countless millions of lives. Now, thirty years after this publication, WHO is back to promoting these methods of malaria control that have never shown any promise in a malaria-endemic region.

Sustainability is a nebulous term that can be used in any context and to advance almost any argument. In this context, it appears that WHO considers sustainable malaria control to be any method that avoids the use of persistent chemicals, even if that necessitates the use of unproven "environmental" methods that could even increase the spread and incidence of the disease. In effect, WHO hides behind this ill-defined concept to satisfy an environmentalist agenda and in so doing, undermines malaria control.

The Stockholm Convention Secretariat, WHO, and environmental groups may well achieve their goals of a DDT-free world, not through the development of any viable

[664] UN News Service, "UN agencies launch DDT-free anti-malaria initiative," May 6, 2009, http://www.un.org/apps/news/story.asp?NewsID=30713&Cr=malaria&Cr1, (accessed Sept. 25, 2009).

[665] WHO Expert Committee on Malaria, 17th Report, WHO Technical Report Series 640, Geneva, 1979, p. 52.

alternative or true replacement, but by the gradual discontinuation of DDT production. Concerns about the ongoing availability of DDT are increasing. Only two countries manufacture DDT, China and India, and although Ethiopia has applied for an exemption for production, it simply re-formulates technical grade DDT imported from China. Most of the DDT used in malaria control in sub-Saharan Africa is sourced from various producers in China and then packaged and re-formulated for IRS in South Africa. Some countries, such as Mozambique, have imported DDT from India's sole manufacturer, Hindustan Insecticides Limited (HIL).

In 2006, the Chinese authorities announced that they would cease production of DDT and other POPs by 2010.[666] Indeed, both China's and India's official exemption for the production of DDT under the Stockholm Convention expired in May 2009.[667] HIL has committed itself to continued production of DDT to meet the needs of the malaria control programs and presumably the Indian government will renew its exemption.[668] However, one producer of DDT may be insufficient to meet global needs; by October 2009, some in the malaria control community were already warning of shortages of DDT for the upcoming spraying seasons. Furthermore, the price of India's DDT has already increased by around 30 percent.[669]

---

666 Xinhua News "Pollutants will be phased out" Shanghai News, June 22, 2006. http://china-environmental-news.blogspot.com/2006_06_01_archive.html (accessed Oct. 12, 2009)

667 Stockholm Convention, "Register of specific exemptions" http://chm.pops.int/Programmes/Exemptions/RegisterofSpecificExemptions/tabid/171/language/en-US/Default.aspx (accessed Oct. 12, 2009)

668 Harikumar, K "Hindustan Insecticides Limited, Chairman's speech" Sept. 24, 2009. http://www.hil-india.com/cmddesk.html (accessed Oct. 12, 2009)

669 Macdonald, M "DDT availability for 2010" Memo for President's Malaria Initiative, Sept. 25, 2009.

In the past, the public health community has responded vigorously to a reduction in the availability of certain public health commodities, such as vaccines and medicines. For instance, numerous advocacy and public health groups responded in 2004 to a shortage of artemisinin for the production of malaria treatments.[670] Shortages of DDT will have equally, if not more, dramatic and harmful effects on malaria cases and deaths and yet there has been a deafening silence from most advocacy groups on this topic. Far from urging countries to continue DDT production, WHO and other UN organizations are more likely to quietly watch DDT supplies dwindle and may even claim this as a public health victory.

DDT is no panacea, but it has a better track record on malaria control than any other intervention. Lives are lost every day because of continued opposition to its use. With development and modernization and, perhaps, a malaria vaccine, DDT will one day no longer be necessary, but that day is still a long way off. In the meantime, people continue to die of a very preventable disease.

Summary

DDT use has always been politically charged and through the use of misinformation and fear tactics it was one of the first major victories of the environmental movement. To this day it remains for them a favorite topic. This continues, even though as described in this chapter it has been used in recent times to protect lives and health of many in South Africa and neighboring countries. There was a tragic malaria epidemic in the late 1990s in South Africa as a result of removing DDT from the routine

670 For example, MSF Press release, "MSF responds to shortage of Novartis malaria drug Coartem announced by the WHO" Nov. 10, 2004. ttp://www.msfaccess.org/media-room/press-releases/press-release-detail/?tx_ttnews%5Btt_news%5D=135&cHash=952f211d68 (accessed Oct. 12, 2009)

malaria control program. That tragedy galvanized the governments in the region to stand firm in defense of DDT and retain it for critical public health use. The Stockholm Convention allows for the use of DDT in malaria control, but ultimately seeks to phase it out. A requirement of the Convention is that parties should provide funding for alternatives, including chemical alternatives. However, as we have explained, this funding has been negligible and has studiously ignored the search for new chemicals.

The successes in controlling malaria through the use of IRS with DDT and other insecticides are reflected in changed policies and strategies at the highest levels within WHO, USAID, and some other donors. These changed, policies, however were not achieved through evidence of success alone, but with concerted advocacy efforts and political interest in malaria control from the U.S. Congress. However, in spite of impressive gains made against malaria through IRS with DDT and other insecticides, environmental advocacy continues to work to undermine these interventions. Vigilance is required to preserve the options for countries to use IRS and insecticides to save lives and protect the health of some of the world's most vulnerable people.

Chapter 8

Malaria—A Disease for the Have Nots

As we wrap up our discussion of DDT and the problems of malaria control, it is worthwhile to review how humanity arrived at the point of hating a chemical that has provided such great benefit in saving lives and reducing disease. The answer is tied to two factors: one is a set of beliefs, and the other is the power of fear as a motivational force.

Externalities from any human activity are a reality, and many environmentalists were well intentioned in wanting to mitigate the potentially harmful effects of DDT spraying. Yet, some within the environmental movement advanced their claims of DDT harm by scaring people into believing that we were inflicting great harm on the environment, wildlife, and our own health when, in fact, we were not. Fear as a sales gimmick is a pivotal tool in the mass media's arsenal of methods to sell newspapers, to keep viewing audiences, and to mobilize the public. So a mutually beneficial alliance sprung to life between the practitioners of fear within the environmental movement and reporters of the mass media. Through this alliance, messages about the evils of DDT were rolled out time and time again.

As we have already described, fear of DDT existed from the very beginning of its use. There was no science behind those claims, but the claims were publicized anyway. It is a fundamental fact that they were publicized not because they were valid, but because they made effective headlines. Media reports of potential harm from DDT rarely put that harm into any kind of perspective or balance, and few journalists report on the results of the many studies that suggest there is no harm, even from high and continuous exposure to DDT. Numerous newspaper headlines about DDT, along with advocacy from anti-insecticide groups, bred suspicion. Suspicion was replaced by fear, and fear was eventually replaced by the conviction that DDT was genuinely harmful and deadly.

The more claims of harm were publicized, the more scientists were stimulated to investigate DDT as a cause of harm. The more they searched, the more they found that DDT was biologically active in a variety of ways (as are all biogenic chemicals), but in each and every case, it showed only weak biological activity. Indeed, this chemical, so broadly denounced as extremely poisonous, is, in fact, hundreds of times less toxic for insects than modern insecticides. And, as we have shown, when it is sprayed in the environment, DDT is degraded, sequestered, or compartmentalized. A chemical that is only weakly active will still be no more than weakly active even if present in the food web. In other words, the small fractions of DDT present in food did not become more dangerous than when originally sprayed on crops. DDT is retained in the food web because of storage in fat cells, and within fat cells DDT becomes less active, not more. Additionally, most living organisms can break DDT down into less toxic components. DDE is one of these breakdown products that are far less acutely toxic than DDT. So, even the presence of DDT in the food chain was weak, with only the smallest of fractions of total DDT found in living organisms. As history has shown, DDT's presence in the food web, even when it was sprayed heavily in the environment, didn't lead to the elimination of organisms or the harms that the tacticians of fear in the environmental movement had predicted.

Life evolved in association with chemicals like DDT, and a plethora of mechanisms for dealing with them evolved as well. Man's production and use of DDT does not constitute a hazard beyond the control of living organisms. DDT sprayed in the environment was applied at doses designed to kill insects. Organisms exposed to those concentrations of freshly sprayed chemical will either succumb or they will neutralize its toxic actions through preexisting mechanisms.

DDT is not just another chemical. It is unique among all the chemicals recommended for malaria control in that it alone functions as a spatial repellent to keep mosquitoes out of houses, provides long-lasting protection, and is cheap to produce and use. Yet, despite decades of field and laboratory observations confirming these unique characteristics of DDT effectiveness, WHO continues to deny them. To this day, WHO emphasizes only killing actions of chemicals for disease control.

Fear tactics eventually produced strong beliefs that DDT was harmful. These beliefs then led to scientifically unwarranted actions, such as the EPA administrator's 1972 DDT ban that was a blatant political act to appease environmentalists. As the EPA itself states:

> Though unpopular among certain segments of EPA's constituency, his [Ruckelshaus's] decision did serve to enhance the activist image he sought to create for the agency, and without prohibitive political cost.[671]

The environmental movement then used its success in achieving the ban to raise funds, which were in turn used to further advance its goals. The ban itself is an amazing and catastrophic display of environmentalism over public health. Some might argue that the EPA administrator could not possibly have foreseen the far-ranging, adverse public-health impacts of the ban. That argument is not valid. Officials from WHO and the Pan American Health Organization (PAHO) spoke in defense of DDT and warned a ban would be disastrous. Even the U.S. Surgeon General warned against a ban. Those within the environmental movement had their day in court during months

[671] EPA, "History - Pesticides and Public Health" available at: http://www.epa.gov/history/publications/formative6.htm (accessed Sept 20, 2009).

of testimony, and under cross-examination, they lost. Furthermore, the full weight of the public-health establishment opposed banning DDT. Regardless, the EPA ignored the many months of testimony and scientific evidence, ignored public-health officials, and even ignored its own administrative judge's carefully considered decision against a ban.

The EPA promoted use of methyl parathion as DDT's primary replacement product. In 1972, the EPA knew, from almost twenty-seven years of widespread DDT usage, that it was not known to cause human deaths or even human illness. It also knew, and openly admitted, that methyl parathion was a documented cause of human deaths. But it was not until twenty-seven years after promoting methyl parathion as a substitute for DDT that the EPA finally came to terms with the risks of methyl parathion use and accepted voluntary cancellation of many registered uses of methyl parathion in 1999 with an admission that the chemical was hazardous to those who worked with it, even when they employed protective clothing and equipment.[672] These dangers, however, were already known and understood at the time of the DDT ban. The EPA was willing to trade an increased risk of poisoning and death for innocent Americans for a victory of environmental activism.

Some might argue that it is well within the EPA's purview to decide to ban DDT because the ban only affects what is done in the United States. They might further argue that the EPA should not be blamed for what happens in other countries because it does not decide what is done there. In fact, the EPA, along with U.S. Department of State and other publicly funded U.S. organizations, has been actively lobbying and formulating policies against DDT in the international arena for many years. As stated on the EPA's website, "The United States has taken a leading role to reduce and/or eliminate certain POPs and their

[672] EPA, "Methyl Parathion Risk Management Decision."

releases on a global basis."[673] The group of "certain POPs" includes DDT. Other than the EPA's role in support of the POPs treaty itself, and all that entails, the EPA and others were leading participants in the work of the North American Commission on Environmental Cooperation that forced Mexico to stop producing and using DDT in its malaria-control program. That success set the stage for the GEF to begin its program for eliminating public-health uses of insecticides within Central America. As a result, many Central American countries now rely on distributions of very large numbers of chloroquine pills for malaria suppression as opposed to the preventive measures of insecticide spraying of house walls. This overreliance on drugs for malaria suppression will eventually lead to more serious malaria problems as parasites become resistant to chloroquine therapy.

The environmental movement has created an ideology against insecticides largely through misrepresentations of science. This type of misrepresentation is typified by a letter from DDT opponents George Woodwell and Paul Ehrlich published January 13, 2005, in the *New York Times*. In this letter, they repeat, for the public record, the environmental dogma that we discussed earlier: "Its [DDT's] chronic use as an insecticide causes the evolution of resistant strains of insects. The resistance is met with heavier applications that simply compound the problems."[674] They are talking specifically about use of DDT for malaria control and their statement is false; but it is a falsehood that is projected with all the authority and status that goes with their positions as elite academicians. As a matter of fact, during all the decades in which DDT has been used, there is no pattern of heavier applications of DDT as a result of

---

673 EPA, "Persistent Organic Pollutants: A Global Issue, A Global Response," last updated Apr. 2, 2009, http://www.epa.gov/oia/toxics/pop.htm (accessed Apr. 15, 2009).

674 Woodwell, G.M. and Ehrlich, P.R., letter to the editor, *New York Times*, Jan. 13, 2005.

malaria mosquitoes becoming resistant to DDT. Resistance normally suggests a need to use a chemical with another mode of action, not a need to use more of a chemical to which the vectors are resistant. Furthermore, as we explained in chapter 3, DDT functions mostly as a repellent, not by killing mosquitoes. For example, based on efficacy studies, India applies DDT for malaria control at one gram per square meter of wall surface, not two grams as recommended by WHO. In spite of DDT resistance and spraying at less than recommended concentrations, India still achieves a desirable level of malaria control where DDT is still used. Mexico has a similar history of success. In spite of DDT resistance, until recent years it continued to achieve a desirable level of malaria control by spraying at recommended concentrations. There is no scientific basis for Woodwell and Ehrlich's statement. Stopping use of DDT, however, has caused millions of malaria deaths and billions of malaria infections. Against this backdrop of devastation, one wonders when false statements like the one by Woodwell and Ehrlich will stop being considered as freedom of expression and start being recognized as contributions toward a humanitarian disaster.

As an ideological movement, environmentalism has become astonishingly powerful, controls vast resources, and is not answerable to any government or to the people of any country. In the United States, the EPA, National Institute of Environmental Health Sciences, and Agency for Toxic Substances and Disease Registry are functional components of this empire. A plethora of non-profit organizations like the Pesticide Action Network (PAN), Greenpeace, WWF, NRDC, and EDF serve as the action arms of these public organizations. The non-profits lobby continuously for DDT elimination as well as for elimination of other insecticides, relying on publicly funded environmental organizations to provide science and support. While the non-profits and governmental organizations contribute tirelessly to DDT elimination, none has made the slightest effort to develop or invest in new chemicals that might be legitimate replacements for DDT. Even the National Institutes of Health has provided almost no research support during the last thirty years for finding a DDT alternative. Vast public resources are used to stop public-health programs and eliminate DDT in malaria-control programs; but nothing is spent to find a truly efficacious replacement. As with the false statements of the environmental movement, governmental attacks on DDT with no investment or support for finding alternatives have contributed to this humanitarian disaster.

In 1981, Thomas Dunlap, a specialist in environmental history at Texas A&M University published an assessment of the EDF actions against DDT, ventured the opinion that "the environmental issue has assumed in some places the aspect of a class war—rich against poor, locals against outside do-gooders."[675] He is referring to the problem of DDT elimination from agriculture, which caused farmers to pay far more for insecticides to produce their crops and increased the prices paid by consumers. All of that came to pass and is in line with Woodwell's solution to the problem of environmental pollution, where "unpopular but increasing restrictions on technology" increase its cost.[676] Yet, Dunlap missed the much larger class war in his analysis. The real war has been against poor people in poor countries. As the environmental movement increasingly leveled its barrage of misinformation against public-health use of DDT, Ehrlich's goal of stopping the "export of death control" to underdeveloped countries largely became reality.

Despite recent policy changes, whether we will move from this point toward malaria and disease control in poor countries or continue down the path of DDT elimination is still up in the air. To be clear, poor people are not the beneficiaries of anti-public-health insecticide advocacy. They are, in reality, targeted by the extremely prejudicial and harmful policies of this advocacy. At

---

675 Dunlap, *DDT*, 235.

676 Woodwell, G.M., "Effects of pollution on the structure and physiology of ecosystems.", p. 432.

the behest of the environmental movement, the policies for eliminating house-spray programs and eliminating DDT are prime examples.

The POPs agreement placed DDT in annex B and stipulated that DDT can continue to be used in public-health programs. Placement of DDT in annex B, as opposed to its being legally banned through placement in annex A, was a result of scientists and concerned citizens around the world banding together to prevent another global catastrophe. Regardless, the POPs agreement stipulates that DDT is to be phased out when appropriate replacements become available. As part of the POPs agreement, the GEF received high levels of funding to help countries phase out use of DDT and thus fulfill the agreement's mandate that DDT will ultimately be phased out of disease-control programs. With generous support from the World Bank and others, the GEF is actively encouraging countries to phase out the use of DDT, even though there is no cost-effective replacement. Of course, no funds were made available or even criteria established for developing cost-effective alternatives. This comprises an extraordinary set of circumstances. The qualification that DDT is to be phased out only when an acceptable replacement chemical is available is just ignored. So the GEF works to eliminate DDT from control programs even though a cost-effective replacement doesn't exist.

Given the totality of the environmental movement's failure to act responsibly and given decades of EPA and UNEP campaigns against basic public-health uses of DDT and other insecticides, we believe it is time to remove from their authority any regulatory control over public-health insecticides. Authority over those precious commodities for protecting people from the great diseases should reside squarely with organizations we hold responsible for disease control. In the United States, that responsibility should reside with the Public Health Service—specifically, the CDC—not the EPA. At the international level, regulatory control over public-health insecticides should reside with WHO, not UNEP or the GEF. The GEF, in particular, should lose its funding and be banned from working to eliminate DDT from malaria-control programs. Additionally progress in the control of malaria will be greatly facilitated by rescinding the anti-insecticides WHA Resolution 50.13 which was adopted in 1997. Public-health agencies would still be lobbied by anti-insecticide groups, however the assigned mission of these agencies is for public health and we could expect that they would be more robust in their defense of DDT.

An important DDT attribute is its persistence. DDT persistence ensures continued protection of people inside houses for months after walls are sprayed. However, with the exception of DDT, the anti-DDT campaign has eliminated all other persistent insecticides (dieldrin, chlordane etc.). The targeting of persistent chemicals for elimination has produced fear within the insecticide industry of any chemical that does not degrade quickly. As a consequence of this fear, no insecticide company is going to invest in a chemical that shows persistence. In any event, there is little profit to be made in public-health insecticides, and from an industry perspective, it is more profitable to produce a chemical that requires repeated application than one that remains and continues to provide protection for months at a time. So, incredulously, the GEF is working in countries around the world to phase out our only chemical that shows persistence, even though there is no cost-effective replacement compound. Worse still, as a consequence, there is little funding, few plans, and hardly any hope that a legitimate replacement is going to be developed. Instead of developing a naturally persistent chemical, the industry is trying to take its commercial products and put them in some slow-release formulation so they will have a longer residual activity. In reality, they are taking already expensive formulations and making them even more expensive in the hope of having a longer residual action.

It is a stunning thought that through decades of effort, environmental organizations spanning the globe have spent billions of dollars, much of this from public sources, attempting unsuccessfully to prove that DDT causes meaningful harm to humans. Meanwhile, ignoring the lack of evidence of actual harm, they dedicate their personnel and financial resources to relentlessly campaigning against DDT. None of the governmental or non-governmental organizations have funded research or made any constructive effort to find a more acceptable replacement chemical. This abysmal record is revealing. It is an exercise of imperious authority over DDT with little concern whatsoever for the adverse and devastating effects of their actions. This amounts to the exercise of authority without responsibility.

Millions of malaria deaths and billions of malaria infections have occurred during more than three decades of environmental activism to stop countries from using DDT. Perhaps the saddest commentary of all is that evidence of highly effective methods of disease control is not enough to counter the enormous investment of the anti-DDT and anti-IRS propaganda. Environmental organizations should be condemned for what they have done, and governments should act to reverse the damage done to malaria- and disease-control programs around the world. Governments should also act to reverse the damage done to the research establishment. Environmentalism has systematically destroyed research that should be working vigorously to develop alternative chemicals that are persistent, cheap, and that function in ways that might, in the long term, replace DDT.

Appendices

Appendix 1

A Brief History of Mosquito-Borne Infectious Diseases

The United States has a long history of infectious diseases and a proud record of finding effective and lasting solutions to them. Yellow fever ravaged much of the United States in decades prior to the 1940s, and so huge were Southern losses from the disease that it became known as the Scourge of the South. It devastated areas ranging from Galveston and Laredo in Texas, to New Orleans (famous as a seat of pestilence), to Mobile, Alabama,[677] to Jacksonville, Florida. St. Joseph, the city that had served as the seat of Florida's Constitutional Convention in 1838–1839, was almost wiped out by a yellow fever epidemic in 1841.[678] By 1876, yellow fever had appeared 741 times in 28 states. The disease affected 228 cities and towns and killed 65,311 people; but the worst was yet to come.[679] In 1878, the epicenter of the epidemic appeared in Memphis, where yellow fever infected 17,600 and killed 5,150. So terrifying was the effect of the disease on Memphis that its citizens almost abandoned the city as a settlement site.[680] By the time the epidemic was over,

---

677 Carroll, J., "Yellow Fever: A Popular Lecture," University of Texas Press, Austin, TX, 1905, p. 13–20.

678 Florida State Parks, "Constitution Convention State Museum," http://funandsun.com/parks/ConstitutionConvention/constitutionconvention.html (accessed Apr. 15, 2009).

679 Grensville, D., *Yellow fever and Malarial Diseases* (Philadelphia: Medical Publication Office, 1876).

680 Ibid.

100,000 people throughout the South had had yellow fever and 20,000 had died from it.[681]

It was not until the turn of the century, when physician and public-health visionary Dr. Walter Reed and his team identified the mosquito, and the particular species, *Ae. aegypti*, as the transmitter, or vector, of yellow fever, that public-health officials could find a way to combat this frightening disease. As explained in chapter 2, in 1898, shortly before Reed's discovery, Dr. Ronald Ross, a British army doctor in India, made another crucial discovery about mosquitoes and disease: that mosquitoes transmitted human malaria. And Giovanni Batista Grassi identified the *Anopheles* mosquito as the specific disease vector.[682] The knowledge that mosquitoes transmitted yellow fever and malaria meant that control efforts for these diseases could be focused on the control or eradication of the vector.

Once public-health officials learned that the *Ae. aegypti* mosquito transmitted yellow fever, they swung into action with remarkable enthusiasm and dispatch. They began their efforts in Havana, Cuba, near where Reed and his team had identified the mosquito culprit. A public-health team headed by William Crawford Gorgas used mosquito control and general sanitation such as removing water containers and eliminating stagnant water to drive *Ae. aegypti* and hence yellow fever completely out of Havana—a city where it had reigned unchecked for

---

Forty five volunteer physicians, mostly from the Northwest, came to Memphis. "One did not remain. The others all contracted yellow fever and thirty of them were dead within a month." Carroll, "Yellow Fever," 21.

The United States was not alone in experiencing the staggering effects of yellow fever. In 1730, a two-month long epidemic caused 22,000 deaths in Europe. Africa and South America faced epidemics through the eighteenth, nineteenth, and twentieth centuries.

682 Desowitz, R.S., *The Malaria Capers: Tales of Parasites and People* (New York: Norton, 1991), 192–98.

generations—in just five months.[683] Impressed by Gorgas' achievement, the U.S. government assigned Gorgas to Panama to see if his team could curb the incidence of yellow fever that had so long plagued canal construction. In 1898, in part because of the lives and money lost to disease, France abandoned its efforts to build the Panama Canal. In the first eight years of the French Panama Canal effort (from 1881 to 1888) there were more than 5,500 deaths from malaria (a death rate of 63.1 per 1000 employees). Gorgas's strategy was to fumigate almost all of the houses in Panama City with sulfur and pyrethrum to eradicate *Ae. aegypti*, and in a mere two years, the disease was eradicated. Gorgas's efforts had a direct bearing on the United States' successful completion of the canal.[684]

Gorgas was understandably proud of his successes, but what excited him most was the possibility that his efforts could not just curb, but actually eradicate, the dreaded yellow fever. In 1909, he enthused,

> Great strides in diminishing the spread of yellow fever all over the world have been made in the last ten years. Ten years ago the four principal centers of yellow fever in the world were Vera Cruz, Havana, Rio de Janeiro and Guayaquil. Of these Havana and Rio de Janeiro were much more important centers of infection than Vera Cruz and Guayaquil. From Havana yellow fever has practically been eliminated, and within the last six months Rio de Janeiro has almost been as successful. In Vera Cruz yellow fever has become sporadic instead of being endemic. Sporadic cases within the last year have

[683] Gorgas, a soldier and medical doctor, was a forceful character and believed very strongly in vector control. Part of his mosquito-control strategy was to get householders to destroy mosquito larvae on their property, or face a $10 fine. See Harrison, *Mosquitoes, malaria and man*, 161.

[684] Spielman and D'Antonio, *Mosquito*, 116–26; Harrison, *Mosquitoes, malaria and man*, 163.

> occurred in several different parts of tropical America, and a few on the west coast of Africa, but nothing comparable with what it was ten years ago. It seems to me that yellow fever will entirely disappear within this generation, and that the next generation will look on yellow fever as an extinct disease having only a historic interest. They will look on the yellow fever parasites as we do on the three toed horse—as an animal that existed in the past, without any possibility if reappearing on the earth at any future time.[685]

An even more ambitious undertaking in yellow-fever eradication was yet to come. In the 1930s, the Rockefeller Foundation decided to fund a massive project named the Cooperative Yellow Fever Service (CYFS) in order to eliminate *Ae. aegypti* from Brazil. They chose an American physician and epidemiologist, Fred Soper, to head the project. The program made use of petroleum Paris Green[686] and pyrethrum sprays against *Ae. aegypti*,[687] the carrier of yellow fever. Coincidentally, *An. gambiae* was discovered in the 1930s to have invaded Brazil from its native range in Africa (presumably imported on cargo ships). This dangerous mosquito began causing severe malaria outbreaks in Brazil. As a consequence, Soper expanded his program to include the eradication of *An. gambiae* from Brazil.[688] Through those eradication efforts, Soper proved that large numbers of workers within a strong and disciplined organization could effectively control or eradicate

[685] Smith, H., "Controlling Yellow Fever," in *Yellow Fever*, ed. George Strode (New York: McGraw Hill, 1951), 555.

[686] Paris Green is the common name for copper acetoarsenite which was once commonly used as an insecticide and animal poison.

[687] Pyrethrum is a natural insecticide derived from the chrysanthemum flower. Before it was used by Westerners it had been used for many generations in the Caucuses region and Northwest Persia.

[688] Tren, R. and Bate, R., "Malaria and the DDT Story," Institute of Economic Affairs, London, 2001, p. 15.

important disease vectors. He and his team worked from eastern seacoast cities toward the interior using traditional larviciding and sanitation measures with good success, although "the use of oil—the best available larvicide—was an expensive and slow procedure" and "at the end of 1939 there were only six states in Brazil and the Federal Capital, which were free from *aegypti*."[689] Oil and Paris Green were used against mosquito larvae, while pyrethrum sprays were used against the adult mosquitoes. Pyrethrum was a quick-acting chemical, but, as its action was short-lived, frequent treatments were necessary, adding greatly to cost. Even with such a massive undertaking, their efforts were falling short. The record of performance of eradication efforts against *An. gambiae* was very different. The contract for the eradication effort and collaboration between the Brazilian Ministry of Health and the Rockefeller Foundation was signed in 1939, and by the end of 1940, the last native-born *An. gambiae* had been eliminated.[690]

DDT Enters the Picture

In 1943, the U.S. military put a cheap, long-lasting, safe, and effective pesticide—DDT—to use against dengue in Hawaii. Dengue epidemics had occurred previously in Hawaii. In 1903, a dengue epidemic had infected 30,000 people, and another multi-year epidemic occurred between 1913 and 1915.[691] The 1943 epidemic infected over 1,500 people before the U.S. Army

---

[689] "Yellow Fever Conference (Washington, DC, Dec 21–22, 1954)," Washington, DC: Pan American Sanitary Bureau, Oct. 1955. Reprinted from *Am J Trop Med Hyg*, vol. 4, no. 4 (July 1955), 611.

[690] Soper, F.L. and Wilson, D.B., "Anopeheles gambiae in Brazil: 1930 to 1940", Rockefeller Foundation, New York City, 1943, pp 262

[691] Tummons, P., "Dengue Outbreak: A Taste of Things to Come?" *Environment Hawai'i*, vol. 12, no. 5 (Nov. 2001), http://www.environment-hawaii.org/archives_summary.php?id=507_0_23_0_C, accessed Apr. 15, 2009.

launched an intensive spraying campaign using DDT, which was a success in terminating the epidemic. Emboldened by this victory, the federal government then funded the eradication of *Ae. aegypti* from Hawaii.[692]

In 1946, the CYFS obtained DDT, which Soper and his colleagues embraced.[693] Sprayed as an emulsion on water containers, one treatment of DDT often eliminated *Ae. aegypti* breeding in an entire village. DDT worked so well that some scientists attending a yellow fever conference issued a cautionary note:

> Considerable progress had been made before the introduction of DDT, but this made *aegypti* eradication very much more rapid and much easier than before. The relative ease with which *aegypti* can be eradicated with DDT caused some of the workers from Brazil to be overly enthusiastic in saying that *aegypti* eradication was easy. In fairness, it was easy in comparison with what went before, but eradication always requires careful administration and a hundred percent coverage of the infested area.[694]

---

[692] This campaign was partially successful. *Ae. aegypti* was eradicated from most of Hawaii, although pockets remained on some islands. Where *Ae. aegypti* was eradicated however, *Ae. albopictus* simply moved in to occupy the niche. Until the small outbreak of 2001, all dengue in Hawaii had been brought back by travelers. Tummons, "Dengue Outbreak."

[693] Soper had worked with DDT ever since the Swiss passed it to the U.S. He conducted typhus trials with it in northern Africa and headed the delousing teams in Naples. "Until DDT there was, essentially, no malaria vector control for the poor peoples of the tropical world. For the first time DDT gave malariologists a weapon to kill a large diversity of anopheline species without need for environmental engineering interventions." Desowitz, *Federal Bodysnatchers*, 77.

[694] "Yellow Fever Conference," Washington, DC, 1954.

Whether it was careful administration, the 100-percent coverage, or just the DDT itself, by the end of 1949, the team's efforts had eradicated *Ae. aegypti* from sixteen states, five territories, and the Federal District of Brazil.[695]

With Brazil's success, other countries also began to use DDT to control *Ae. aegypti* and to curb yellow fever epidemics. Iquique, Chile, added DDT to water at a rate of 1 ppm and in so doing reduced the *Ae. aegypti* population to 0. DDT was also extensively used in Caracas, Venezuela, when a case of infectious yellow fever entered the country. Emergency measures sprayed all houses within 400 meters of where the case occurred with residual DDT and the city was also sprayed from the air with DDT.[696]

While Gorgas was successfully, if expensively, battling mosquito-borne diseases in the tropics,[697] malaria was still a major public-health threat in Europe. In 1906, there were over 230,000 cases of malaria in Italy and an estimated 4,871 deaths.[698] In sharp contrast to the focus on vector control by U.S.-funded malaria-control efforts, however, the Italian approach focused on malaria as a social problem and primarily used drugs for malaria control and prevention. Under the influence of renowned public-health expert Angelo Celli, the Italian government concentrated its early twentieth-century malaria-control efforts on improving access to quinine—the first-line treatment for the disease. Apart from the efforts to

---

[695] Ibid., 605, 606, and 613.

[696] Ibid.

[697] Gorgas spent over $350,000 (around $8.2 million in 2008 dollars) per year on disease control and employed over 1,200 health workers.

[698] A case fatality rate of 2.43 percent is now considered very high. (The South African Dept of Health aims to keep its case fatality rate below 1 percent.) The number of malaria cases and deaths are likely to have been grossly underreported however; Bruce-Chwatt and Zulueta comment that the number of cases at the end of the nineteenth century could have been as high as 2 million out of a population of 30 million.

ensure wider access to quinine, the Italians also attempted to drain swamps and relocate people away from marshy areas. While these early Italian efforts cut malaria deaths quite successfully,[699] there was no concomitant dramatic decline in cases (see figure A1-1). During and shortly after World War I, both malaria cases and deaths increased sharply, possibly because of a shortage of quinine, and possibly because of the global influenza epidemic, which may have compromised any partial immunity that patients normally would have had.[700]

Figure A1-1: Malaria Cases and Deaths in Italy, 1887–1978

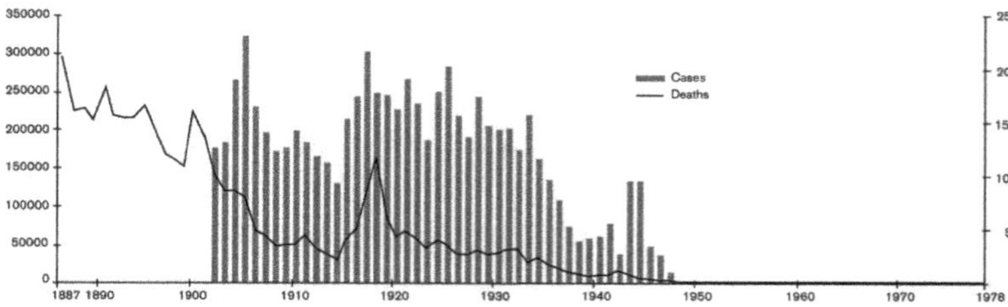

Source: adapted from Bruce-Chwatt and Zulueta (1980)[701]

As a result of changes to the Italian countryside and a way of life that was independent of any malaria-control interventions, malaria cases were steadily falling. For instance, as Professor Paul Reiter explains, from the late nineteenth century, as farmers

---

699 After quinine became more widely available in Italy after a legislative act in 1900, malaria fatalities fell dramatically.

700 Perhaps some influenza deaths were recorded as malaria deaths as well—a practice that occurs today in many poor countries that lack proper diagnostic tools or resources.

701 Data from Prof. Alberto Coluzzi, Institute of Parasitology, University of Rome, published in Bruce-Chwatt, L. and Zulueta J., *The Rise and Fall of Malaria in Europe: A historic-epidemiological study* (Oxford University Press, 1980), 96. Although data on malaria mortality were collected from 1887 onward, it was only after the Italian government promulgated a number of laws in 1901 that made it compulsory to declare malaria cases, hence the lack of data on cases from 1887 to 1902.

developed more land for agriculture, the breeding sites of the *Anopheles* mosquitoes disappeared.[702] Improved and new root crops meant that farmers had sufficient winter fodder to keep livestock all year round. Increased wealth meant that farmers kept their livestock in barns away from their houses. Since the prevalent *Anopheles* mosquitoes feed on both humans and animals, the change in livestock housing and the overall increase in livestock meant that the probability of transmitting malaria between humans was reduced. This is because the more non-human meals that an *Anopheles* mosquito has, the fewer human meals it has and the lower the chances that it will transmit malaria parasites from one human to another.[703] Moreover, as agriculture became more mechanized, fewer laborers were required. The population of rural areas dropped and the cities began to grow. Malaria is primarily a disease of rural populations, so as these populations declined, so too did malaria. Also, people increasingly lived in houses with features that improved protection from mosquitoes, such as glass windows.

The first large-scale DDT application project was implemented on the Italian island of Sardinia in 1946. In four years, the incidence of malaria dropped from 10,000 cases to four.[704] Elated at the results, public-health authorities began

---

702 Reiter, P., "Climate change and mosquito-borne diseases," *Environ Health Persp*, vol. 109, supplement 1 (Mar. 2001). The organized land reclamations during the 1920s and '30s, and those that occurred naturally as part of economic development, were important in reducing malaria cases. However, this form of malaria control appeared not to sustain low malaria transmission as, particularly in times of crisis such as World War II, malaria cases reached epidemic proportions again.

703 Having greater numbers of livestock kept in barns away from humans also contributed to the decline in malaria in the upper Mississippi Valley in the United States. The benefits of sleeping away from livestock are still important, as those at risk from avian flu will attest.

704 Spielman and D'Antonio, *Mosquito*, 50. Although malaria was eliminated, the mosquitoes continue to live and thrive on the island.

dreaming of a worldwide campaign to eradicate malaria through the application of DDT.

Malaria in the United States

In the United States, yellow fever was the most dramatic of the mosquito-borne diseases and the most spectacularly lethal, but the less lethal and far more common malaria had the greatest impact. Vivax malaria, caused by *P. vivax* parasites, was endemic to the Americas and at one point ranged over the entire United States east of the Rockies with substantial pockets in California as well. Slavery transported the more deadly falciparum malaria, caused by *P. falciparum*, out of Africa and into the southern United States. In 1946, estimates of malaria's impact ranged from a conservative 1 million cases to a more speculative 6 million cases a year, although all parties seemed to agree on a death rate of about 5,000 a year.[705] Malaria's economic impact was also staggering. The direct cost of sickness was as much as $51 million (or over $589 million in 2009 dollars), whereas the overall economic impact ranged as high as

[705] Ackerknecht, E., *Malaria in the Upper Mississippi Valley, 1760–1900*, (New York: Arno Press, 1977) (original copyright 1945, JHU Press), 1. See also Leary, J.C., Fishbein, W.I., and Salter, L.C., *DDT and the Insect Problem* (New York: McGraw-Hill, 1946) (estimate of 4 million cases a year and 4,000 deaths). The reason for the vast disparity in the estimate of the number of cases was that, even in the United States, good data was difficult to obtain. The U.S. Public Health Service (USPHS) depended upon physician response to annual surveys, surveys that had an extremely low response rate. But the USPHS authors argued that despite this low rate the surveys gave a good overview of the incidence of malaria in the state. "The results of circularization … is dependable and furnishes quite satisfactory information regarding the geographic distribution of the disease." Malaria in North Carolina" (reprint no. 411) July 20, 1917, p. 4–5. Despite the millions of differences in the estimates of incidence, the death rate estimate remained relatively constant, due to a reliance on mortality reports.

$500 million a year (equivalent to over $5.4 billion in 2009 dollars).[706]

In the United States, malaria's range was extensive, but it had the most impact in the South. An English saying of the time accurately captures malaria's grip on the region: "Those who wish to die go to Carolina." Malaria fares best in warm, moist regions and worst in cooler areas. So as swamps were drained, land developed, and living standards and housing improved, malaria began to disappear from the more temperate regions of New England during the eighteenth century. The Southern states were much less fortunate. The more moist environments there, combined with longer mosquito-breeding seasons enabled by more months of warm temperatures, a lack of interest in draining swamps and marshes, and the slave trade's introduction of new parasites, strengthened malaria's grip on the South.[707] Endemic malaria, epidemic yellow fever, murine typhus,[708] and other ailments in the region gave the South the worst health in the nation and caused Northern life insurance companies to charge higher premiums to Southern policy-holders. In fact, New Orleans "was popularly known as 'the graveyard of the Southwest' because of its frightful mortality rate (nearly three times that of Philadelphia and New York)."[709]

These conditions persisted throughout the century. Although in the decade before the Civil War malaria had started to decline, it was still the chief cause of debility and loss of economic efficiency in the South at the time of the war.[710] The war, however, reversed the trend of improved living standards and

---

[706] Ackerknecht, Malaria in the Upper Mississippi Valley, 1.

[707] O. Breeden, J., "Disease as a Factor in Southern Distinctiveness," in *Disease and Distinctiveness in the American South*, Todd L. Savitt and James Harvey Young, eds. (Knoxville: Univ. Tenn. Press, 1988), 9.

[708] CDC Report. "CDC Activities, 1946/47" US Public Health Service, Communicable Disease Center, Atlanta, Georgia. p. 11

[709] Ibid., 10.

[710] Ibid., 11.

declining malaria cases. After the war, malaria exploded across the South, reaching record levels of infection and spreading back into areas where it had declined. By the time of the Depression in the 1930s, despite twenty years of malaria-control efforts in the South, malaria still had not been brought under control.[711]

The more common vivax malaria did not have a high lethality compared to yellow fever, the chief epidemic disease of the South, but because of the nature of malarial infection, it is not an exaggeration to say that malaria kept the American South poor.[712] Unlike yellow fever, which confers long-term immunity to survivors of the disease, malaria can re-infect victims many times. Over time, malarial victims do develop a partial immunity to the disease, which makes their symptoms less severe, but they still experience weakness and enervation from the disease sufficient to prevent them from functioning at normal levels of human capacity.[713] As malaria was more prevalent in rural, agricultural areas and because it flourished during the summer and fall - the most active times in the agricultural year - malaria had a particularly high impact in the agricultural South. A United States Department of Agriculture study of malarial prevalence on a "well-managed plantation in Louisiana" discovered that "from May to October [there was] a total loss of time due to malaria sickness of 1,066 days in 74 tenant families—60 families would have done the same amount of work had there been no malaria. The actual loss of time

711 Dr. Rose Carter, H., Testimony before Congress, "Hearing on Malaria and Typhoid," 1918.

712 Dr. Barrow, C., "Transactions of the Third Annual Conference of Malaria Field Workers. Proceedings of Third Annual Conference of Malaria Field Workers," Hot Springs, Arkansas, Nov. 15–17, 1921 (Washington, DC: GPO, 1922).

713 Subcommittee on Engineering and the USPHS, "Malaria Control for Engineers," National Malaria Committee, Washington, DC, 1936, p. 1–2; Carter, "Hearing on Typhoid and Malaria," p. 12–13.

amounted to $2,200 and the loss of efficiency to $4,300, an amount of over $63,000 when inflated to 2005 dollars."[714]

It was because of the profound economic impact of malaria that early malaria-control crusaders' chief allies were not local governments, but rather private businesses that lost so much to the effects of malaria.[715] One early control effort focused on a saw mill in North Carolina with stunning success.[716] Once local governments saw the positive effect that malaria control had on private operations, they became more inclined to support malaria-control efforts in their communities. Likewise, when the federal government saw the positive effects of malaria control, it became more interested in supporting control efforts.

The history of malaria and other insect-borne diseases in the U.S. and Europe shows that this disease once prevailed in temperate zones and is not limited to tropical climes. With increasing wealth and development and with the right set of tools, these temperate zones beat malaria and other insect-borne diseases such as yellow fever and murine typhus. Other countries can achieve similar gains against these diseases with the right set of tools, such as DDT and other effective insecticides. However, securing a set of public institutions, such as good government, the rule of law, and property rights, that allow individuals and societies to prosper all need to be in place to create the bedrock upon which sustainable and lasting public health programs can be built. Much has been made about the need for sustainable malaria control programs. Perhaps the most critical need is for sustained funding for disease control from donor agencies and consistent, science-based leadership from the UN agencies that set the global public health agenda. These agencies have proven to be highly vulnerable to political

---

[714] National Malaria Committee, 1936, p. 2.

[715] Dr. Rose Carter, H., Third Annual Conference of Malaria Field Workers, p. 3; Subcommittee on Engineering and the USPHS, 1936, p. 3, 38.

[716] Carter, Third Annual Conference of Malaria Field Workers.

pressure and, like weather vanes, have withdrawn support for highly effective interventions and backed less effective ones based on prevailing political winds. The capricious nature of policies emanating from these agencies has done much to undermine and weaken previously strong and effective national disease control programs.

Appendix 2

The Bermuda Petrel

The petrel (*Pterodroma cahow*) is the national bird of Bermuda.[717] As BirdLife International, a global partnership of conservation organizations, explains, the bird

> once bred abundantly throughout Bermuda (to UK). It was thought extinct for almost three centuries, until reported (with specimens) during the first half of the 20th century. In 1951, 18 pairs were rediscovered breeding on suboptimal rocky islets (total area 1 ha) in Castle Harbour. Intensive management has resulted in slow but steady increases, and the population was estimated at 250 birds in 2005, with 70 pairs fledging a record 40 young in 2003, and 71 pairs fledging 35 young in 2005.[718]

The report goes on to say that the bird

> formerly nested in soil burrows, but such habitat is not available on current breeding islands and it now nests in suboptimal, natural erosion limestone crevices and artificial burrows. The breeding season is January–June, and breeding success has increased from less than 5%

[717] BirdLife International, "Bermuda Petrel–BirdLife Species Factsheet," http://www.birdlife.org/datazone/species/index.html?action=SpcHTMDetails.asp&sid=3910&m=0 (accessed Apr. 15, 2009).

[718] Ibid.

> per year in the 1950s to more than 25% per year in the 1990s.[719]

Before Bermuda was colonized, the petrel was reportedly "superabundant." It is "a slow breeder, but excellent flier, and spends its adult life on the open seas."[720] It returns to its former nesting place after five years and "begins breeding, laying only one egg per season."[721] The birds mate for life.

Reportedly, the bird's "eerie nocturnal cries stopped the early Spanish seafarers settling the Islands out of superstition, as they thought the Isles were inhabited by Devils. Instead they put ashore hogs as a living foodstore [*sic*] for passing ships."[722] Bermuda was, however, colonized by the British, and their introductions of non-native animals like rats, cats, and dogs, along with mass killings by the early colonists, practically eliminated the petrel.

BirdLife International specifies the true threats that brought the Bermuda petrel to the brink of extinction: "The drastic population decline is attributed to habitat loss, exploitation and predation. Its recovery has been hampered by competition from White-tailed Tropicbird *Phaethon lepturus* for nest-sites."[723]

Those who brought the petrel back from brink of extinction were true heroes. As the BirdLife International website describes, their work was hard and entirely directed at the true threats to the bird's existence.

---

[719] Ibid.

[720] National Audubon Society, "Bermuda petrel, pterodroma cahow," http://web1.audubon.org/science/species/watchlist/profile.php?speciesCode=berpet (Accessed Sept 27, 2009)

[721] Ibid.

[722] Animal Pictures Archive, "Bermuda petrel, pterodroma cahow," http://www.animalpicturesarchive.com/view.php?tid=3&did=28576 (Accessed Sept 27, 2009)

[723] Birdlife International, "Bermuda petrel"

Since 1961, there has been management of nesting-grounds, including the creation of artificial burrows, periodic removal of rats and the elimination of nest-site competition from *P. lepturus* (by installing baffles over burrow entrances). As part of the Bermuda Conservation Programme, potential breeding islands (e.g. Nonsuch) have been reforested with native flora in an attempt to attract nesting petrels. The Castle Harbour islands are a National Park and Nature Reserve. In 2004 and 2005 efforts were made to attract adult Cahows [Bermuda petrels], displaced from low-lying nest burrows destroyed by hurricane Fabian, on the main breeding islet to a new artificial burrow complex built on a more elevated section of the islet. Using a combination of a sound attraction system set up among the new nests and physicall [*sic*] translocation of adult Cahows from the destroyed sites, three pairs occupied burrows in the new complex by March, 2005. A project is now underway to establish a new nesting colony on the Nonsuch Island Living Museum, which is much larger and contains more suitable breeding habitat than the present suboptimal breeding islets. The project involves physical translocation of chicks from the present breeding islets to a new complex of artificial burrows on Nonsuch, so that they will imprint on the new site and return when mature to establish their nests at the new location. In 2004, the trial year of the project took place with 14 chicks moved to Nonsuch, where they were fed and monitored every other day until departure, with all fledging successfully. In 2005, 21 chicks were translocated, with all again fledging successfully by mid-June. This project is scheduled to continue for three more years, with a target of 90 to 100 chicks in total being translocated over a five-year period.[724]

[724] Ibid.

Described here is the real work of wildlife-management and species-recovery efforts. It is not related to the false claims of anti-insecticide environmentalists. Indeed, such claims rob credit from the real heroes of species recovery. Missing from descriptions of work on behalf of the Bermuda petrel is any mention of DDT. The high-visibility claim of the Environmental Defense Fund that DDT was causing extinction of the Bermuda petrel was pure fiction.

Appendix 3

Chemical Threats to Wildlife

Birds fed high doses of DDT might experience changes in eggshells. Other than these high dose-effects, DDT has not been shown to cause reproductive harm. As it turns out, other chemicals also cause eggshell thinning, such as mercury and certain polychlorinated biphenyls (PCBs). Furthermore, not uncommonly, the other compounds can be embryo-toxic and more abundant than DDT. In addition, many of these compounds are natural products, not man-made. Most of these chemicals have gone largely unnoticed by the popular media because of continuous efforts to keep attention riveted on DDT. Some who were originally involved in making a case against DDT struggle to deflect attention from those other chemicals. This is illustrated in a publication by Robert Risebrough, an ecologist, who adheres to his claim that DDT, not PCBs, was responsible for eggshell thinning. But science marches on and his claims are widely refuted by scientists in many countries working on PCBs.

PCBs were widely used from the early part of the twentieth century as coolants and lubricants in transformers and other electrical equipment and, because of this use, they are found widely in the environment. Tests in 1975 showed that selected types of PCBs could reduce the viability of chicken eggs:

> Hatchability of fertile eggs as a result of biphenyl feeding was significantly affected by 10 or 20 p.p.m. and by three PCB compounds (Aroclors 1232, 1242, 1248). The hatchability (%) of eggs laid by pullets fed 10 and 20 p.p.m. declined to 78 and 45, respectively, for

Aroclor 1232, 48 and 5 for Aroclor 1242 and 40 and 5 for Aroclor 1248 after 6 weeks of biphenyl feeding, as compared with 90 for the control group.[725]

As science and technology improved, it became possible to study many other persistent chemicals and more specific types of compounds that were, in many instances, more abundant and more toxic than DDT. Some have been shown to have adverse affects on wildlife. For example, populations of red kites (large birds of prey) in Donana National Park in Spain have reportedly declined by more than 50 percent in the last five years. A survey of egg contaminants suggested that one cause of population decline was PCBs, stating that "seventy-five percent of the red kite eggs had PCB levels above 4.7 microg/g f.w., which is associated in the literature with reproductive failure."[726]

As evidence against DDT being the sole or even primary culprit in wildlife endangerment accumulates, the debate has spun beyond the control of those who created the anti-DDT hullabaloo in the first place. Even Theodora Colborn, a well-known environmentalist, reveals some of the dimensions of this controversy:

> Evidence appears sufficient to state that DDT and its metabolites are the causal agents [of harm] until one takes a closer look. Postupalsky (1971) noted that eggshell thinning was 9% of the problem in inland Wisconsin eggs. As mentioned earlier, Koeman et al. (1972) suggested that something other than eggshell thinning was the cause of embryo mortality in white-

[725] Lillie, R.J., et al., "Toxicity of certain polychlorinated and polybrominated biphenyls on reproductive efficiency of caged chickens," *Poult Sci*, vol. 54, no. 5 (1975): 1550–55.

[726] Gomara, B. and Gonzalez, M.J., "Enantiomeric fractions and congener specific determination of polychlorinated biphenyls in eggs of predatory birds from Donana National Park (Spain)," *Chemosphere*, vol. 63, no. 4 (2006): 662–69.

tailed sea eagles; uncracked eggs were experiencing 30–50% mortality also.[727]

Colborn also cites other studies countering the claim that DDT was the sole culprit. On the subject of other chemicals, she states that "PCB appears to be related to unexplained population declines in a number of papers seeking a causal agent . . . PCBs and similarly acting dioxins and furans (Birnbaum et al., 1985; Bradlaw et al., 1980) cannot be ruled out."[728]

A recent study by wildlife toxicologists John Elliott and Ross Norstrom on bald eagle productivity on Canada's Pacific coast revealed reduced productivity in areas of pulp mills, a common source of various dioxins.[729] Dioxins, as the FDA explains, "are formed as a result of combustion processes such as commercial or municipal waste incineration and from burning fuels (like wood, coal or oil)."[730] Likewise, the great blue heron colonies in the vicinity of pulp mills have shown reproductive failures as high as 71 percent. The authors of the heron study concluded that "the predominant factors influencing reproductive success were probably disturbance by humans and bald eagles, combined with loss and degradation of nesting habitat, and not sub lethal toxicity."[731] As we see in the results of

[727] Colborn, "Epidemiology of Great Lakes Bald Eagles," 430.

[728] Ibid., 432.

[729] Elliott, J.E. and Norstrom, R.J., "Chlorinated hydrocarbon contaminants and productivity of bald eagle populations on the Pacific coast of Canada," *Environ Tox and Chem*, vol. 17, no. 6 (1998): 1142–53.

[730] Interagency Working Group on Dioxins, "Questions and answers about dioxins," available from U.S. Food and Drug Administration, Center for Food Safety and Applied Nutrition, http://www.cfsan.fda.gov/~lrd/dioxinqa.html (accessed Apr. 13, 2009).

[731] Harris, M.L., et al., "Reproductive success and chlorinated hydrocarbon contamination of resident great blue herons (*Ardea herodias*) from coastal British Columbia, Canada, 1977 to 2000," *Environ Pollut*, vol. 121, no. 2 (2003): 207–27.

this study, in spite of a complex of chemical contaminants, the real cause of heron endangerment relates to far more obvious and direct sources of harm. As explained in the main text, for the bald eagle and peregrine falcon, the real causes of bird endangerment generally rest with those growing sources of harm, such as hunting and land-use changes, that were reducing their populations decades before the advent of DDT.

As an aside, Harris and co-authors' conclusion that sublethal toxicity was not the cause of harm to the great blue heron colonies is noteworthy. Although this report is not one of a kind, it is notable in that the authors were willing to look beyond the narrow focus of a typical chemical contaminants study. All too often, investigators of ecotox studies show little interest in attributing wildlife harm to any cause other than an insidious effect of chemical contamination. In the course of this study, Harris et al. downplay the nebulous concept of sublethal toxicity and all of its claims of harm that are almost never proven but, once claimed, are almost impossible to disprove.

There are many biological and non-biological causes of reduced wildlife productivity. These include human disturbance, malnutrition, population pressures, diseases, predation, chemical pollutants and adverse weather. When one examines the history of population declines for specific wildlife species, as illustrated in the main text, it becomes evident that environmental contamination with synthetic chemicals has been the least of those causes. However, these many causes were abundantly present during all the years when science and scientists, as well as the public and elected officials, were being manipulated into blaming DDT. The motivation behind this manipulation was three pronged. Some believed that DDT was saving lives and allowing for unacceptable rates of population growth. Others were simply opposed to the application of science and technology and others honestly, but erroneously, believed that insecticides were causing wildlife harm. For example, the great diversity of persistent, natural chemicals is illustrated by the dioxins:

> "Dioxins" refers to a group of chemical compounds that share certain chemical structures and biological characteristics. Several hundred of these compounds exist and are members of three closely related families: the chlorinated dibenzo-*p*-dioxins (CDDs), chlorinated dibenzofurans (CDFs) and certain polychlorinated biphenyls (PCBs). Sometimes the term dioxin is also used to refer to the most studied and one of the most toxic dioxins, 2,3,7,8-tetrachlorodibenzo-*p*-dioxin (TCDD). CDDs and CDFs are not created intentionally, but are produced inadvertently by a number of human activities. Natural processes also produce CDDs and CDFs. PCBs are manufactured products, but they are no longer produced in the United States.[732]

As Suter explains, there are 209 theoretically possible types of PCBs and over 210 other dioxin-like compounds.[733] The dioxins are just one example of the many chemicals that are more abundant, more persistent, and in some cases more biologically active and toxic than DDT or DDT metabolites. Many of these chemicals are more abundant in the food chain than DDT and, like DDT, some were more abundant in the past than they are today. As an example, consider the dioxin PCB. The discovery of the PCB problem in 1966 is explained by an environmental advocacy group, the Environmental Working Group (EWG), in its descriptions of PCB contamination in waterways in Anniston, Alabama.[734]

---

[732] Suter II, G., "Analyses of laboratory and field studies of reproductive toxicity in birds exposed to dioxin-like compounds for use in ecological risk assessment," Apr. 2003, EPA/600/R-03/114F.

[733] Ibid.

[734] Chemical Industry Archives, A project of Environmental Working Group, "Anniston AL, in depth. Monstanto responds to PCBs in wildlife," http://www.chemicalindustryarchives.org/dirtysecrets/annistonindepth/indepth.pdf (accessed Apr. 13, 2009).

> In 1966, the world got its first glimpse into the persistence of PCBs when a Swedish scientist who was studying the pesticide DDT (another Monsanto product) found a second compound that was exhibiting many of the same characteristics in tests. The "mystery peak" on the scientists' testing equipment turned out to be PCBs.[735]

According to EWG advocates, in the 1960s, Monsanto was producing PCBs at its plant in Alabama and PCBs were included in plant outflow into a local stream called Snow Creek. The effect on stream life was reportedly alarming. Tests involving caging of fish in the contaminated drainage system of Monsanto's plant were used to observe the effect of the contaminants. According to documents produced by EWG,

> all 25 fish lost equilibrium and turned on their sides in 10 seconds and <u>all were dead in 3 minutes.</u> The gill covers (opercles) immediately assumed a flared position, and blood issued from the gills after 3-minutes exposure . . . Our preliminary inspections indicate Snow Creek to be devoid of life . . .
>
> The outflow . . . contains some extremely toxic materials and kills fish in less than 24 hours when diluted 300 times. In a flowing system (as opposed to our static tests) and under conditions of constant exposure, this effluent would probably kill fish when diluted 1000 times or so.[736]

According to the EWG, an internal communication showed that Monsanto was worried about PCBs becoming involved in the environmental movement's drive against insecticides. The basis for their concerns relates to the broad chemical similarities of DDT and PCBs. EWG advocates document one Monsanto

[735] Ibid.

[736] Ibid.

official as saying, "The wildlife people are dedicated to the demise of DDT."[737] The quote is not entirely correct in grouping those who oppose DDT as wildlife people, although some were. Many of the early leaders in the anti-DDT movement had almost no credentials as wildlife people, toxicologists, or environmental epidemiologists, and most had no noteworthy track record of DDT research. They were, first and foremost, environmental activists masquerading as experts. For example, a PubMed search of the names of each of the six authors of an important anti-DDT paper published in *Science* magazine in October 1970 revealed that not one of them had any other peer-reviewed paper on DDT.[738]

What has been described in this short appendix is a plethora of chemicals that are, in one way or another, similar to DDT. Beyond the ones mentioned here, there is an abundance of natural chemicals (known and unknown) that have similar characteristics (see chapter 5). However, while the natural chemicals are ignored, the man-made chemicals with similar characteristics have been under broad attack by environmental activists for the last several decades. The activists have employed scary scenarios and frightening claims to grab public attention and motivate political action against those chemicals.

Overall, there are certain consistencies in environmental propaganda against man-made chemicals targeted for elimination. First, the focus of claims of harm is invariably on a nebulous but insidious effect. An insidious effect is, from the activists' perspective, a highly desirable claim in that it is almost impossible to disprove. Another consistency in the propaganda is to ignore the need to apply adequate standards for proof of a cause-effect relationship.

The broadly accepted epidemiological criteria for cause-effect relationships have previously been described as follows:

---

737 Ibid.

738 Harrison, H.L., et al., "Systems studies of DDT transport," *Science*, vol. 170, no. 3957: 503–508

> In order to prove a cause-and-effect relationship, certain core criteria should be met, such as strength of association, biologic credibility, and consistency with other investigations. Two other important, although sometimes considered weaker, criteria are time sequence (cause must precede effect) and a proportional dose–response relationship.

In disease epidemiology, a claim of a cause-effect relationship is not acceptable unless the relationship fulfills these criteria. In contrast, among those studies of chemical exposures causing wildlife or human health harm, it is very difficult to discern a comparable set of criteria for cause-effect relationships. Another important consistency is that claims of wildlife harm have, with few exceptions, ignored a plethora of non-chemical factors that have historically accounted for devastating harm to wildlife populations. We have reviewed in detail many of the non-chemical factors in the main text on the bald eagle, peregrine falcon, deer, wild turkey, American bison, heath hen, and the passenger pigeon.

Last but not least, there is a consistent failure in eco-toxicological claims of DDT causing human health harm to show any improvement at all in human health once DDT has been eliminated from disease-control programs. In contrast, we find ample evidence of serious increases in human disease and significant losses of human health once use of DDT has been stopped. Establishing robust cause-effect relationships is important in the scientific process but becomes absolutely critical when decisions based on these relationships could impose enormous costs to human health and lives. The history of activism against DDT based on weak or non-existent cause-effect relationships has been disastrous, costing lives without any discernable benefit.

Appendix 4

DDT as a Cause of X-Disease

Starting in the late 1940s, when it became clear that a limited quantity of DDT was stored in human body fat, people have linked DDT to an enormous variety of ills. The lack of proof for these linkages did not seem to disturb them. In 1956, Leonard Wickenden's *Our Daily Poison: The Effects of DDT, Fluorides, Hormones and Other Chemicals on Modern Man* linked DDT to polio, mysterious but dramatic fatal illnesses, viral diseases, and particularly to X-disease (hyperkeratosis) in cattle and hepatitis in humans.

In 1939, veterinarians started noticing a new disease in cattle which they termed X-disease. In this disease, the cattle's epidermis, the outermost layer of the skin, became inflamed and thickened. Along with the changes in the epidermis, the cattle lost weight, developed wart-like swellings in the mouth, drooled, and had runny noses. The disease seemed to spread from state to state and caused considerable losses to cattle farmers.

X-disease continued to affect cattle across the United States until the early 1950s when researchers at the School of Agriculture at Virginia Tech discovered the source of the malady. When visiting a farm afflicted with the disease, a veterinarian from the school, Dr. Wilson Bell, noticed that the cattle had access to the full barnyard, including access to the farm machinery. He further noticed that the cattle would lick the axles of the farm machinery. Bell obtained from the farmer a sample of the lubricant he used and fed that to calves back at Virginia Tech. The calves developed X-disease.[739]

Bell and his team then investigated the lubricant and discovered that the company that made it, Lubrizol Corporation of Cleveland, Ohio, used a particular additive in the product, chlorinated naphthalene.[740] The agricultural chemistry lab at Virginia Tech tested the compound extensively on rats and calves and found that it did, in fact, cause the disease.[741] By 1954, the U.S. Department of Agriculture (USDA) was able to publish a pamphlet that stated that chlorinated naphthalene was the cause of X-disease.[742]

In 1949, however, a doctor named Morton Biskind had published several articles in a variety of publications that linked DDT to a variety of ills, including X-disease in cattle. Biskind's articles never offered any actual proof for his assertions. Unlike the research teams at Virginia Tech, he did not run laboratory tests to see if feeding DDT to cattle caused X-disease as feeding chlorinated naphthalene to cattle had, nor did he cite any such

[739] Virginia Tech, "Oral History interview with Ruben W. Engel," http://spec.lib.vt.edu/archives/depthistory/biochem/oralhist/engel/engel1.html (accessed Apr. 15, 2009).

[740] Ibid.

[741] "Woodrow and I did an awful lot of work with rats, and so forth, and so on, and calves. When we finally found out what the problem was, we had to dose some calves with hexofluronapthalene. Their eyeballs would actually pop out on the ground, you know. That stuff was potent….It was a grease—old farmers would buy a 100 pound drum and use it for ten years, maybe. And the cattle would come along and lick it off. It didn't take much. You could inject a little in rat and just a few days later his liver was about twice as big as it ought to be. We did that, and went through a bunch of rats." Virginia Tech, "Oral History interview with Juel Albert," http://spec.lib.vt.edu/archives/depthistory/biochem/oralhist/albert/albert1.html (accessed Apr. 15, 2009).

[742] United States Dept. of Agriculture, *Hyperkeratosis (X-disease) of cattle*, USDA leaflet no. 355 (Washington, DC: GPO, 1954).

studies. Instead, he simply implied that these illnesses were caused by DDT exposure.[743] Wickenden picked up where Biskind left off and went one step further, rejecting the proof that chlorinated naphthalene causes X-disease in cattle. He sneered that the notion is impossible, as cows wouldn't lick machinery or drink crankcase oil,[744] and he urged the reader to focus on the far more common chlorinated hydrocarbon, DDT, as the likely cause of the hyperkeratosis.

Biskind and Wickenden's work illustrates the ways in which people commonly skewered DDT with causing all the ills of the world. The reports of harm are anecdotal at best, levied through specious innuendo at worst, and are shaped around a truth: some new chemicals did have detrimental health effects. But to shape a statement around a truth does not mean that the statement is the truth. To say that some chemicals have detrimental health

---

[743] Biskind admitted as much when he testified before Congress in 1950. "I suggested for purposes of investigation that the same etiology may apply to 'X disease' of cattle, which has caused tremendous losses of livestock in this country in recent years. Here too I do not know and have never stated that the original outbreaks of this type in 1939 were caused by DDT poisoning—I have no way of knowing whether DDT was available in this country for experimental or other use a year after its introduction as an insecticide in Switzerland." Morton S. Biskind, "Statement On Clinical Intoxication From DDT And Other New Insecticides," Presented before the Select Committee to Investigate the Use of Chemicals in Food Products, United States House of Representatives, Dec. 12, 1950, http://www.vaccines.plus.com/Bishkind-1951-DDT.html (accessed Apr. 15, 2009).

[744] Mr. Wickenden was obviously as ignorant about cows as he was about DDT. Cows will lick anything that seems vaguely interesting and will eat almost anything that crosses their paths including batteries and even airplanes. See BBC News, "Cows like a light aeroplane snack," http://news.bbc.co.uk/2/hi/uk_news/england/west_midlands/3614022.stm (accessed Apr. 15, 2009). They will also drink crankcase oil. See Anon, "Lead Poisoning in Cattle," The Government of Alberta, 2001, http://www1.agric.gov.ab.ca/$department/deptdocs.nsf/all/agdex791 (accessed Apr. 15, 2009).

effects on humans, DDT is a chemical, and therefore, DDT has a detrimental health effect on humans is an existential fallacy in the absence of proof. And yet this argument has been made time and time again in the case of DDT, most famously in Rachel Carson's *Silent Spring*.

Carson insisted that man's use of insecticides was an insult to nature: "Control of nature is a phrase conceived in arrogance, born of the Neanderthal age of biology and philosophy." Of course, nature's balance includes a vast diversity of its own chemicals, including biocides, repellents, attractants, and deterrents. Living organisms use their own insecticides on a scale and range that scientists are still investigating, and have been using these insecticides since long before man tried to replicate them. Carson, however, believed that chemicals had no role in the natural order of life, and despite its scientific illiteracy, this view is now common.

Appendix 5

Litigation Against DDT

On December 29, 1969, a suit was filed against the use of DDT in the United States Court of Appeals. A *New York Times* article described the case.

> Four conservation groups and six individuals have taken their case against DDT to the United States Court of Appeals here . . . This petition was filed by the Environmental Defense Fund, Inc., the Sierra Club, the West Michigan Environmental Action Council and the National Audubon Society. It asked the court to review the Department of Agriculture's actions in light of the earlier petition asking for a complete ban on DDT . . . The other petition filed today was submitted by the Environmental Defense Fund and six individuals, all of them residents of California. It asked for a review of the Department of Health, Education and Welfare's rejection of an earlier petition on DDT traces in foods . . . In effect, this petition had asked that no traces of DDT be allowed in foods . . . The petitions directed at the Department of Health, Education and Welfare cited evidence that DDT had been found capable of causing cancers in mice that were fed large doses of the chemical. The petitioners are hoping to get DDT traces in food banned.[745]

---

[745] *New York Times*, "EDF takes DDT appeals to court," 1969. Reprint of *The New York Times* article about early Environmental Defense Fund litigation on DDT, http://www.edf.org/article.cfm?contentID=2447 (accessed Apr. 15, 2009).

Environmental activists and EDF lawyers were often successful in state and local court cases where they orchestrated local support. In those settings, the defense of DDT was not well-supported, prepared, or organized. The machinations of the EDF and those who supported the EDF in litigation against DDT are described in considerable detail in Thomas R. Dunlap's 1981 book, *DDT: Scientists, Citizens, and Public Policy.*[746] Dunlap, who was clearly an ardent fan of the EDF's work, provides considerable insight into the organization's premeditated attempts to grab headlines and push forward with a national campaign against DDT and all other insecticides. As Dunlap explains, the EDF's first case in Wisconsin was to stop the use of DDT for Dutch elm disease. EDF partnered as legal counsel to the Citizens' Natural Resources Association (CNRA), which filed a complaint with the Wisconsin Department of Natural Resources and named as defendants the city of Milwaukee and Buckley Tree Service, which it alleged had a contract with the city to spray the elms. Maurice Van Susteren, chief hearing examiner for the department, assigned himself to the case.

The hearing began with an off-the-record discussion that effectively closed the case. The city and Buckley agreed not to use DDT and showed that there was no contract between them. The EDF then asked for a three-day adjournment to prepare an amended complaint, and a six-day hearing in which to present the full case. The hearing examiner denied the request because the issue was moot. The EDF side was distraught and some of the environmental campaigners were reduced to tears. According to Dunlap, "Van Sustern remarked to the clerk that these people were crazy. They had won, what did they have to be unhappy about?"[747] The EDF lawyer explained that winning was not the goal—they wanted a public forum

[746] Dunlap, *DDT*, 152.

[747] Ibid., 153.

> "before whom to present their scientific evidence and get a judgment. Van Susteren pointed out that if that was the case, they had chosen the wrong legal route."[748]

That early hearing showed the EDF's true intent—they wanted to make headlines in order to maintain the public record of their trumped-up charges against DDT. As described in a history of those proceedings, from that first case, the EDF was guided by Van Susteren along a path for achieving its goals in Wisconsin.

> Wisconsin law provided a declaratory ruling procedure that was much more suitable. Any Wisconsin citizen could ask a government department for a ruling on the applicability of a particular set of facts to any rule enforced by the department. The concerned agency would then hold a public hearing in which each side in the dispute would present its case, subject to rules of evidence and cross-examination. CNRA members, for example, could ask the Department of Natural Resources if DDT was a water pollutant under the Wisconsin water-quality standards. The department would then hold a hearing in which the EDF could attempt to show that DDT did, in fact, contaminate the waters of the state and injure the human or animal life of the area.[749]

This was the avenue the EDF was searching for and, acting as legal representatives of CNRA, it immediately launched a suit against DDT in Wisconsin.[750] The CNRA was joined in this case by the Wisconsin Division Izaak Walton League of America and the Michigan Audubon Society. The EDF's lawyer, Victor

---

[748] Ibid.

[749] Ibid.

[750] Ibid., 154.

Yannacone, was lead counsel. The National Audubon Society, Charles Wurster, and the Wisconsin Resource Conservation Council, among others, threw their support behind the petition. Buckley Tree Service, the Wisconsin Arborist Association, and Egelhoff Tree Service were the primary local organizations in opposition. Several others were listed as opposing the petition.[751] The Wisconsin petition requested a declaratory ruling that "DDT is a highly toxic persistent chemical, that its use be restricted in such a way that it cannot enter the biosphere and that its existence in the biosphere constitutes pollution."[752]

The Examiner's Report defined a major obstacle in the hearing: "Without doubt DDT has provided enormous economic benefits, but economic benefits are not an issue or part of any issue in this case." Of course, this structuring of the petition to exclude considerations of DDT's benefits was entirely contradictory to Senator Nelson's opening comments to the court and the audience: "The specific question before us is whether the overall benefits of DDT are offset by the damage it does."[753] By not examining economic benefits, the hearing was already skewed against DDT.

As described by Dunlap, "The hearing, in effect, allowed it [the EDF] to choose its ground, set the terms of discussion, and force the industry to come and defend itself."[754] As it turned out, the Industry Task Force for DDT of the National Agricultural Chemicals Association did not find out about the hearing until a week before it started,[755] so the opposition to the EDF's actions came into the hearing unprepared. Thus, those listed in the petition as DDT supporters had already declined to fight for

---

751 University of Wisconsin Marathon Country, "Wisconsin DDT Hearing Examiner's Report, 1970," http://www.uwmc.uwc.edu/geography/350/DDT-hearing.htm. (accessed Sept 25, 2009)

752 Ibid.

753 Iltis, "Population prophet."

754 Dunlap, *DDT*, 159.

755 Ibid., 159.

continued use of DDT. On the other hand, the organization that was truly interested in defending DDT (the Industry Task Force for DDT) was not listed as a party to the petition and had a belated entry into the proceedings.

While the defenders of DDT were poorly organized, a group of local scientists and University of Wisconsin staff and students provided enormous assistance to the EDF and the case against DDT. As Dunlap explains:

> Hickey [an ornithologist at the University of Wisconsin and long term anti-DDT activist] asked William Reeder, a member of the zoology department, to help prepare information on the scientific background for the environmentalists. Reeder, initially detached, was soon caught up in the enthusiasm and recruited his own and other graduate students (including some in economic entomology), to do the library work. During the hearing he had a place at the counsel table, and whenever someone mentioned a paper, quoted a scientist, or brought up a new line of questioning, he would send a messenger out to call his assistants at the university. They in turn would look up the article, find out the scientist's background and qualifications, or assemble a group of papers on the topic under discussion. Usually Yannacone had a fund of information within an hour.[756]

With such considerable resources attacking DDT and without any consideration of DDT's actual benefits versus its theoretical harms, the outcome was a foregone conclusion:

> DDT and its analogs are therefore environmental pollutants within the definitions of Sections 144.01 (11) and 144.30 (9), Wisconsin Statutes by contaminating and rendering unclean and impure the air, land and

[756] Ibid., 156

> waters of the state and making the same injurious to public health and deleterious to fish, bird and animal life.[757]

It was, of course, this type of careful structuring of petitions, formulations of charges, astute legal work, and management of the media that made the difference in the outcomes of the EDF suits. Examples of headlines at that time were "DDT Uncontrollable, Scientist Says" in the *Milwaukee Journal* and "DDT Termed Peril to the Sex Organs" in the *New York Times*.[758] The Wisconsin suit was ideological and lawyers for the task force knew it. One exchange between the task force and the EDF lawyers emphasizes this point. Louis McLean, leading the taskforce defending DDT, said,

> I think it will become apparent, Your Honor, that while this particular action is limited to DDT, that the activities of this organization [the EDF] in the last few years will indicate that they are not simply against one chemical, but any, the use of any pesticides.
>
> Yannacone: That's a pile of hogwash.[759]

In contradiction to Yannacone's statement, the EDF had, in fact, already filed suits against dieldrin, and by the end of 1970, the EDF had also gone after aldrin, another organochlorine widely used in agriculture. As Dunlap reports, just two years after that, "the EDF was working against almost all the persistent chlorinated pesticides on the market, seeking both cancellation and suspension."[760] So, the task force lawyer's claim was hardly hogwash and was prescient of a pattern

---

[757] University of Wisconsin Marathon Country, "Wisconsin DDT Hearing Examiner's Report, 1970."

[758] Dunlap, *DDT*, 166, 174.

[759] Dunlap, *DDT*, 165.

[760] Ibid., 237.

environmentalists would follow: opposing all insecticides while denying that they oppose any and all uses of insecticides. Opposing all insecticides would discredit environmental campaigns, making them seem radical and unreasonable. Attacking one chemical at a time, however, gives the appearance of a more reasoned and scientific approach.

On May 21, 1970, after a year of hearings, the ruling of the Wisconsin case banning DDT's use in Wisconsin was published.

The claims that the EDF, the Audubon Society, and other environmental groups made during the lawsuit were eagerly reported as headline news. In their continuing effort to scare the public and propel political processes against DDT, opponents took advantage of this media visibility and started to advance the myth that DDT in breast milk would cause cancer in babies. To push this argument, the Audubon Society created and passed out anti-DDT bumper stickers that had a skull and crossbones on them.[761] These campaigns helped to make DDT a topic of national concern.

What happened in 1969 is perhaps the best example of what can be achieved through a campaign of fear. It was a pivotal year when a series of critical actions were taken against DDT. In the fall, the Nixon administration ordered the cancellation of all residential uses of DDT even though the United States' *Ae. aegypti* eradication program, which had been initiated in 1964, was entirely dependent on DDT. By backing out of the hemisphere-wide program started in 1948 to eradicate this important disease vector, the United States effectively pulled the rug out from under successful eradication programs throughout the Western hemisphere. Countries that had eradicated *Ae. aegypti* needed the United States, which was a major trading partner of those countries, to eradicate the species to prevent re-introductions from the U.S. Once the U.S. cancelled its eradication program, maintaining the *Ae. aegypti*-free status in

[761] Jukes, T., "DDT Stands Trial Again," *BioScience*, vol. 11 (Nov. 1972):670–676, 670.

other countries became hopeless. By the 1980s it became apparent that for all of Central and South America, a growing epidemic of dengue fever was the consequence of that congressional action. But the rush of actions against DDT was continuous. Sweden also banned most uses of DDT in 1969,[762] and this was also the year that the World Health Organization ended the global malaria-eradication program, which was also entirely dependent on DDT.

President Nixon promised to eliminate all but the essential (including selected agricultural pests and its use in public health) uses of DDT by the end of 1970.[763] He then announced to Congress in February 1970, "We have taken action to phase out the use of DDT and other hard pesticides."[764] Environmental groups, however, thought the administration's action was insufficient. They wanted a complete ban on DDT and, in October 1969, the EDF filed a petition with the Secretary of Agriculture requesting the immediate suspension and ultimate cancellation of all uses of DDT. The Secretary refused in June 1970, stating that "the use of DDT should continue to be reduced in an orderly, practicable manner which will not deprive mankind of uses which are essential to the public health and welfare."[765] The EDF responded by petitioning the D.C. Court of Appeals for review.

---

[762] Wahlstrom, B., "Managing POPs in Sweden." Swedish National Chemicals Inspectorate, http://www.chem.unep.ch/pops/POPs_Inc/proceedings/bangkok/WAHLS1.html (accessed Apr. 15, 2009).

[763] Flippen, Nixon and the Environment, 44.

[764] Hooper, J., "A Study of the Environmental Protection Agency Decision to Cancel Use of DDT Using the Kennedy Behavioral Diagnostic Schematic," Northern Illinois University, May 1975, 15.

[765] Statement of the Reasons Underlying the Decisions on Behalf of the Secretary with respect to the Registration of Products Containing DDT, June 29, 1970, at 13 cited in *EDF v. Ruckelshaus*, 439 F. 2d 584, 594 (1971).

In December 1970, a new player entered the field when Congress created the Environmental Protection Agency (EPA) and conferred upon it the authority to regulate pesticides. Thus, in January 1971 when the D.C. Court of Appeals responded to the EDF's petition by questioning the safety of DDT and ordering an administrative hearing,[766] the hearing was conducted not by the Department of Agriculture, but by the EPA. With the EPA hearing, the EDF's brief but successful history of court verdicts in its favor seemed like it would end when the EPA's hearing examiner, Edmund Sweeney, ruled that DDT was not harmful to humans or wildlife and that there was a need for its continued use (discussed in chapter 5). Two months later, however, EPA administrator William Ruckelshaus overturned the decision in a political move and canceled all uses of DDT for crop production and non-health purposes.

In its attempt to get DDT banned in the United States, the EDF and its allies argued that DDT was a carcinogen. While the EDF appeared content to focus only on the status of DDT in the United States, the Audubon Society, one of the EDF's allies, went even further, urging its supporters to join in the campaign for banning the export of DDT. This development and the carcinogenic charge greatly alarmed WHO. It feared that these claims would have a detrimental effect on the anti-malaria programs in developing countries.[767] Despite the fact that WHO, PAHO, and even the U.S. Surgeon General defended DDT and warned of the dangers in banning it, the momentum appeared to be with the DDT opponents. Even the head of the White House Council of Environmental Quality supported an immediate ban, asserting that DDT and other pesticides killed over two hundred people per year.[768]

So, against a background of hysteria that had been created by the fear tactics of the environmental movement, a full

[766] EDF v. Ruckelshaus, 1971.

[767] Jukes, "DDT Stands Trial Again."

[768] Flippen, Nixon and the Environment, 140–41.

administrative hearing was conducted. It was the first major environmental case against a registered insecticide under the Federal Insecticide, Fungicide and Rodenticide Act (FIFRA). The EDF and the EPA composed the team against DDT.[769] Their chief witnesses on the effects of DDT on wildlife were Robert Risebrough, George Woodwell, and Joseph Hickey. Charles Wurster, chairman of the EDF's scientific committee, was to be the anchor witness. Samuel Epstein was their chief witness on human-health factors.

Dr. Gordon Edwards' analyses and literature citations,[770] insights from Robert Ackerly, the chief trial counsel for the DDT industry,[771] and Dunlap's book, *DDT: Scientists, Citizens, and Public Policy*,[772] provide a deep well of information on the willingness of the EDF and the EPA to put truth and science aside to achieve a DDT ban.[773] Consider the testimony of Robert Risebrough, an ecologist from the University of California, Berkeley. In cross-examination, Risebrough presented completely contradictory information about brown pelicans on the Channel Islands off the coast of California. He stated that there were 500 birds in 1969. In earlier trial testimony in other

[769] In context of the administrative hearing one might think the EPA would take a neutral position and not argue decisively against DDT; but one would be wrong. As stated by Ackerly, "From the opening moment of the pre-trial proceedings, however, the EPA attorneys assumed a wholly adversary stance and crusaded vigorously to prosecute and secure the [DDT] cancellation." Quoted from Ackerly, Robert L. DDT. A re-evaluation. Part I. Chem Times and Trends (Oct 1981) 47-53, 51.

[770] Edwards, J.G., "DDT: A case study in scientific fraud," *J Amer Phys & Surgeons*, vol. 9, no. 3 (2004): 83–88.

[771] Ackerly, R.L., "DDT A re-evaluation. Part I," *Chemical Times and Trends* (Oct. 1981): 47–53.

[772] Dunlap, *DDT.*

[773] Dunlap's book gives a very well researched and detailed history of the EDF campaign against DDT though its tone is somewhat biased towards EDF and against DDT.

cases, he had testified that the pelican was extinct or "gone" from the Channel Islands. His arguments about DDT and eggshell thinning as a result of DDE, a metabolite of DDT, were unconvincing. In fact, it was shown that DDE levels were higher in pelicans in Florida than in California and that populations in Florida were reproducing normally.[774]

The DDT defense team put data on trends of raptor populations compiled at Hawk Mountain Sanctuary into the record. Dr. Hickey, the EPA witness, attempted to discredit the significance of the Hawk Mountain counts. He later had to admit that he had used those counts as a credible source of population data in his own publications. In contrast to Hickey's testimony, the trend data showed that the osprey and the peregrine falcon populations were not in decline. For example, as revealed by Ackerly, data for the osprey showed

> 31 counted in 1934 to 530 in 1969. The addition of another observation site at Hawk Mountain in 1967 might account for some of the increase. But a review shows that the osprey count was 191 in 1946 and reached a pre-1966 high of 444 in 1965.
>
> The peregrine falcon, which showed a low count of 14 in 1935 and 1965, has ranged since 1934 in the 20's and 30's, with counts of 22 in 1966, 22 in 1967, 21 in 1968 and 32 in 1969.[775]

Actual peregrine and osprey population-trend data did not support EDF and EPA claims, so the witnesses had no recourse but to try to minimize the significance of the Hawk Mountain counts. EPA and EDF arguments fell flat under the cross-examination of their experts. As summed up by Ackerly,

---

[774] Ackerly, "DDT: A re-evaluation. Part I."

[775] Ibid.

the final test of the theory that DDT residues in the environment have adversely influenced bird reproduction must be the condition of the bird populations in that environment. Since EPA and EDF contended that DDT residues were rather uniformly wide-spread [which, of course, they are not], the validity of their theory should have been reflected uniformly among species of birds and consistently in bird population surveys. The data of record, such as the Hawk Mountain surveys, not only did not support their theoretical concepts, but were compelling evidence of the lack of validity of the thesis.[776]

George Woodwell was presented as an EDF expert on the fate of DDT in the environment,[777] and yet a modern search of peer-reviewed literature reveals only four publications by Woodwell on DDT: one on DDT in forest soil,[778] one on DDT and fiddler crabs,[779] one on DDT in an estuary,[780] and one on DDT in the biosphere.[781] Woodwell's characterization of environmental pollution problems was presented in a 1970 paper in which he offered his solution to the pollution problems: "fewer people" and "unpopular but increasing restrictions on

[776] Ibid.

[777] Ackerly, R.L., "DDT: A re-evaluation. Part II," *Chemical Times and Trends* (Jan. 1982): 48–55.

[778] Woodwell, G.M. and Martin, F.T., "Persistence of DDT in soils of heavily sprayed forest stands," *Science*, vol. 145 (1964): 481–483.

[779] Odum, W.E., Woodwell, G.M., Wurster, C.F., "DDT residues absorbed from organic detritus by fiddler crabs," Science, vol. 164 (3879), May 2, 1969: 576-577

[780] Woodwell et al., "DDT residues in an East Coast Estuary."

[781] Woodwell et al., "DDT in the biosphere."

technology."[782] It is perhaps no coincidence that all four of these papers were published in *Science*.

When Woodwell gave testimony on January 12, 1972, he made no mention of the "DDT in the biosphere" paper. This paper was discovered by the industry and USDA lawyers after the trial and admitted into the record by the EPA administrator.[783]

> Woodwell . . . wrote (Science 174:1101, 1971) that "six billion pounds of DDT had been used, but only 12 million pounds could be accounted for in all the earth's plants, animals, fish and birds." Which (he stated) was "less than a thirtieth of one year's production of DDT during the mid-60's." His testimony at the EPA Hearings was grossly different than this, so reporters asked why his data in Science were not mentioned during his testimony. Woodwell responded that EPA lawyers told him not to mention the information in that article, "lest my testimony be disallowed." (Business Week, 8 July 1972).[784]

In Ackerly's opinion, the paper by Woodwell, Craig, and Johnson revealed findings that, if the paper had been revealed, would have harmed the EPA and EDF case against DDT.

In this paper, Woodwell undertook to tabulate from all known data sources the amount of DDT present in all of the biota of the world. He stated, "We consider this an estimate of the maximum amount of DDT that could be in the biota" and concluded, "the total quantities of DDT residues in the biota are

---

782 Woodwell, G.M., "Effects of pollution on the structure and physiology of ecosystems." *Science*, 20 April 1970, 168:429-433, p. 432

783 Ackerly, "DDT: A re-evaluation. Part II".

784 Edwards, J.G. "Science, pesticides and environmentalist politics" Executive Intelligence Review, December 10, 1000 pp 20-31. (page 9).

but 1/30 or less of the annual amount of DDT produced in recent years." Woodwell explained why the theoretical prediction that DDT residues remain in large quantities in the environment was not valid:

> The physical and chemical characteristics of DDT might lead one to assume that the biosphere should behave as a giant separatory funnel, gradually partitioning the lipid soluble residues into the lipid rich biota. Although there is no question that this process does occur, there is also no escape from the conclusion that it does not work well on the biospheric level. *Most of the DDT produced has either been degraded to innocuousness or sequestered in places where it is not freely available to the biota* (emphasis added by Ackerly).[785]

In his analysis, Ackerly described another study that confirmed limited quantities of DDT in the biota, lack of DDT persistence in the biosphere, and lack of biomagnification of DDT in the food chain.[786] The study was conducted at the Woods Hole Oceanographic Institute. The results, as described by Ackerly, were that "DDT is not as persistent as once believed and is confined fairly closely to the point of its introduction."[787] EPA research has shown

> that the half-life of DDT in naturally illuminated seawater is about ten days. Consequently, DDT is now undetectable in Atlantic deep-sea water or sediments . . . Another unexpected result of our data and that of others was a lack of support for the concept of food chain magnification. There are now sufficient data from many sources in the scientific literature to conclude that food

[785] Ackerly, "DDT: A re-evaluation. Part II."

[786] Ibid.

[787] Ibid., 49.

> chain magnification of pollutants is not a useful concept with waterbreathing animals.[788]

The findings described by Ackerly for the Woods Hole Oceanographic Institute are consistent with observations we presented earlier about the natural world's ability to deal effectively and efficiently with DDT as both a risk to the environment and a risk to wildlife.

Ackerly did not mention, and perhaps did not know, that Woodwell's model had already been severely criticized within the scientific literature. A letter to *Science*,[789] published in 1971 by the chemist Clare Stewart, Jr., concluded that the amount of DDT in the air, based on Woodwell's cited source for his calculations, was perhaps a mere 1/10 of Woodwell's estimate. Stewart stated that Woodwell's model, as presented, required many assumptions, and that "the validity of their conclusions must be questioned on the basis that measurements of atmospheric concentrations of DDT, as judged by the one reference given [by Woodwell], are completely incompatible with the requirements of the model."[790] Stewart went on to conclude that Woodwell was wrong in his conclusion that "the accumulation of DDT with continued usage is 'causing spectacular declines in populations' of various species. If there is serious factual evidence that this is so, it would be appropriate for the authors to cite it separately from the many items of speculation, suggestion, and opinion which were provided."[791]

Woodwell also presented testimony on a study he co-authored with Charles Wurster, who was to be the anchor witness for the EPA and EDF's claims. The objective of the study was to measure the accumulation of DDT in the biota of a

---

[788] Ibid.

[789] Stewart, C.A., "Atmospheric circulation of DDT," *Science*, vol. 177: 724–25.

[790] Ibid.

[791] Ibid.

marsh and the adjacent estuary. Furthermore, the study was designed to investigate how DDT residues accumulate in the sediment of the marsh. As we detail in the full extract from Ackerly's publication in the footnote, Woodwell and Wurster claimed that DDT accumulated to a concentration of thirteen pounds per acre, even though it was only used at concentrations of one pound per acre in the surrounding fields. However, on cross-examination, Woodwell revealed that the samples were collected in six areas on the edge of the marsh where spray equipment was cleaned; this would inevitably increase the concentration of DDT at those points. In addition, Woodwell conceded that, in fact, with more comprehensive sampling, the concentration was only one pound per acre. Woodwell never published a correction or retraction of his claim that DDT had concentrated to a level of thirteen pounds per acre, reasoning, "I never felt that this was necessary."[792]

---

[792] Ackerly, "DDT: A re-evaluation. Part II," 50–51. Ackerly reports on this incident as follows: "Dr. Woodwell explained the background of this paper, stating the point of the paper to be: 'The question was the extent to which residues had accumulated in that marsh and in the biota of the marsh, in the estuary adjacent to it.' He stated that a primary conclusion reached was: 'The second point that we made in this paper was that residues accumulate in the sediments of the marsh, particularly in the organic sediments of the spartina marsh, to high levels.' " Ackerly goes on to say, "The paper reports finding a concentration of DDT in the marsh of 13 pounds per acre. Since the rate of application was one pound per acre this represents a dramatic buildup of residues. The paper has been cited repeatedly by fellow scientists to support the proposition that annual applications of one pound per acre can build up to a level of 13 pounds per acre. On cross-examination, Dr. Woodwell conceded that the 13 pounds per acre was based upon six samples along the edge of the marsh where the spray equipment was cleaned. He also stated that upon more extensive sampling, the overall figures reflected a concentration of closer to 1 pound per acre. The 13 pounds per acre has never been corrected in the literature, nor has the follow-up study ever been published. When asked whether he had ever published a retraction or correction of the 13 pounds per acre concentration, Dr. Woodwell answered: 'I never

After cross-examining Woodwell, industry lawyers wanted to cross-examine the co-author of the estuary study, Charles Wurster. EDF counsel announced that Wurster, who had been described as the anchor witness, had decided not to testify. The judge demanded that Wurster appear for testimony, but he did not.

Edwards also described Woodwell's testimony and drew attention to the false information:

> George Woodwell's EPA testimony (p. 7232, 13 January 1972) referred to Science 156:821 (1967) where the abstract by Woodwell and Wurster stated "DDT residues in an extensive salt marsh on the south shore of Long Island average more than 13 pounds of DDT per acre." Under cross-examination, Woodwell admitted that his reference to 13 pounds of DDT per acre should actually have been "an average of about one pound per acre." Also he said they had only taken six samples, and that "our initial sampling was in a convenient place for the mosquito commission's spray truck, too."[793]

It was this level of error in experimental design and misrepresentation of scientific findings that Edwards found objectionable. Edwards spent much of his life defending DDT and attempting to ensure accurate, scientific analysis of the chemical.[794]

---

felt that this was necessary.' (Tr. 7238.) This was just part of the cross-examination prepared by industry lawyers for Dr. Woodwell's co-author, Dr. Wurster. The day before Dr. Wurster's scheduled appearance, counsel for EDF announced that he had decided not to testify on the ground that his testimony would be cumulative and thus not necessary."

[793] Edwards, "DDT: A case study in scientific fraud."

[794] He did not shy away from visual aids to get his point across about DDT's safety for humans and reportedly ate DDT while lecturing on the subject. Edwards suffered no ill effect from his DDT consumption.

The question of whether DDT was a human carcinogen was addressed during the EPA hearings. Claims against DDT had already circulated heavily in the press and public concern had been aroused. This was a critical issue to the outcome of the trial, yet little testimony was actually given on the subject. The U.S. Surgeon General, Jesse L. Steinfeld, M.D., and Wayland J. Hayes, M.D., Ph.D., a U.S. Public Health Service scientist, both testified that DDT had not been shown to cause cancer. Dr. Samuel Epstein, who had conducted studies with mice exposed to DDT, contended that DDT did cause cancer. Epstein's position was based on an incomplete and inconsistent study in which mice were first fed 100 ppm and then 300 ppm of DDT. Epstein then went on to recommend that as an alternative to DDT, methoxychlor be used. Methoxychlor, however is known to be more toxic to humans than DDT.[795]

He died of a sudden heart attack on July 19, 2004, at age 85 while mountaineering in Glacier National Park.

[795] Ackerly, "DDT: A re-evaluation. Part II," 50, gives a full account of Epstein's paper and his cross examination. Little testimony was devoted to the issue of whether DDT was a human carcinogen (see Part I, p. 48, CT&T, Oct. 1961).

Only Dr. Samuel Epstein, who is credited by the Administrator with merely a footnote reference, contended that DDT poses a high risk of being a carcinogen in human beings. Dr. Epstein introduced into the record an incomplete summary of an unpublished mouse study which he contended was significant in determining whether DDT, in fact, causes liver tumors in mice.

Through a variety of techniques, including requests under the Freedom of Information Act to the Department of Health and Human Services, as well as to the WPA, the attorneys discovered that Dr. Epstein submitted only a portion of the study; that the results were erratic that in some instances the control animals developed more liver tumors than those fed with DDT and then, most remarkably, that the investigators lost control of the study and instead of feeding 100 ppm of DDT to the animals, as the summary suggested, the animals were fed 300 ppm for a substantial portion of the study. (Pleadings File No. I-50, Mar. 16, 1972.)

Edwards discussed Epstein's testimony in a 2002 paper that he sent to one of this book's authors (Roberts). In a note of transmittal, he said he had distributed about fifty copies of the document. Edwards carefully cited sources in the document, but

---

This glaring inconsistency between opinion and fact is, unfortunately, the common occurrence in that record. Relying heavily on a 1969 unpublished paper called the "Fitzhugh Study," Dr. Epstein concluded that DDT posed a clear hazard to man. In addition, he suggested that methoxychlor could replace DDT.

Q. And yet you're willing to make a statement like, "Methoxychlor can replace the persistent and carcinogenic DDT."

A. This refers to the fact that methoxychlor is not persistent. There is no clear evidence on carcinogenicity of methoxychlor. There is some evidence that methoxychlor can be used as an alternative to DDT, as evidenced by the reference which I cite in my article.

Q. You don't say "some evidence," Doctor. You say, "Methyoxychlor can replace DDT." (Tr. 7386).

The complete study showed a malignant tumor incidence in the methoxychlor-treated mice which was greater than the incidence for the DDT-treated mice. It is strange indeed that Dr. Epstein did not investigate the full "Fitzhugh Study" before considering it "definitive evidence" that DDT is carcinogenic for man and before posing mehtoxychlor as a substitute for DDT.

Dr. Epstein's posture as a crusader was clearly revealed during this cross-examination. He attributed ulterior motives to scientists who testified contrary to his point of view (Tr. 65: 7361) and severely criticized a prestigious medical journal of the American Medical Association for publishing a paper with which he disagreed (he also cited the wrong journal) (Tr. 65-7363). He implied that industry might be withholding studies and reports when he had not the slightest basis for this accusation (Tr. 65:7390.).

The "Fitzhugh Study" was so completely and thoroughly demolished and discredited that the Administrator did not refer to it in his Order. He did, however, refer to Dr. Epstein's testimony, the effectiveness of which was based upon incomplete reporting of this defective study."

it remains an unpublished work. Edwards included observations on research that fully refuted Epstein's testimony. These observations help create a deeper understanding of why the administrative judge, Edmund Sweeney, decided that DDT should not be banned. Edwards' comments about Epstein's testimony and favorable coverage of Epstein's testimony by *Science* are particularly enlightening. It is worth noting that an authentic DDT expert, Dr. Wayland Hayes, presented comprehensive testimony at the EPA hearing, yet *Science* opted to give "Point of View" coverage to the only witness who incessantly claimed DDT was a cause of human cancer.

A modern PubMed search of publications on DDT will produce a list of at least nineteen publications by Hayes and two by Epstein. One of the two Epstein papers is just a response to Dr. Paul Russell's criticism of the *Science* coverage of Epstein's testimony.[796] Russell was the leading malaria expert in the United States at the time of the EPA hearing, and he objected to Epstein's views. The second Epstein paper is on dieldrin, not DDT. Edwards' comments are as follows:

> Samuel Epstein testified at the EPA hearings that "DDT causes cancer." He cited as his source an unpublished study by Fitzhugh, Gross and Davis. (Gross had stated that the report was not published because they discovered that the mice had been fed 300 mgs/kg.) Epstein testified that 68 of the DDT-fed mice developed cancer, but avoided mentioning that 83 of the "control" mice (<u>with no DDT</u>) also developed cancer. (The researchers called them "tumors" but Epstein referred to them as "cancers"). Epstein's omission of the significant data was fraudulent, as was the reference to "cancers" instead of "tumors". Science magazine favored Epstein's fraudulent statements, and even published a

[796] Epstein, S.S., "DDT toxicity," *Science*, vol. 177, no. 47 (1972): 388; Russell, P.F., "DDT toxicity," *Science*, vol. 177, no. 47 (1972): 387–88.

summary of his faulty testimony, in 1972 [Science, Vol. 175, page 610]. In response to criticisms for publishing that misinformation, editor Nicholas Wade informed me that: "We chose to print it because it seemed to us to be outspoken and controversial. We will doubtless return to the subject when the hearings are finished, and Dr. Epstein's testimony will then be placed in its proper perspective." (They did not "return to the subject," and Epstein's fraudulent testimony was never discredited in Science.) Incidentally, Dr. Epstein stated under oath during the hearings (p. 7306) that he was member of the Health, Education and Welfare Panel on carcinogens, but under cross-examination (page 7374) he later admitted that he was not.

But has DDT caused cancer, anywhere? Obviously not. There have been NO reports indicating cancers resulting from DDT inhalation or ingestion. In fact, the opposite results have been documented. In numerous studies DDT has been found to prevent cancer.[797]

Edwards' numerous scientific references are given in the footnote below.[798]

---

[797] Edwards, "DDT: A case study in scientific fraud," 12–16.

[798] HEW (Mrak) Report (of the secretary's Commission on Pesticides and Environmental Health (1969), "No reliable study has revealed a causal association between the presence of DDT residues and human diseases."

Hayes et al, "Evidence of Safety of Long-term High Oral Doses of DDT for Man." "After daily doses of 35 mgs of DDT for 21 months there was no definite chemical or laboratory evidence of injury by DDT."

Fitzhugh, O.G. and Nelson, A.A., *J Pharmacol Exper Therapy*, vol. 89 (1947): 18–30. Their rats did not develop cancer, even when fed 1,000 ppm DDT in their food for twelve weeks or 100 ppm DDT for two years. They also found no liver changes in chickens, dogs, cats, or monkeys fed DDT. The director (Lehman) of FDFA summarized it by concluding: "DDT is not a carcinogen."

Ottoboni, A., *California's Health*, May 1972. "No tumors were formed in four generations of dogs (about 500 dogs) fed 1,000 to 10,000 times the daily intake by Americans."

Laws, E.R., et al., *Arch Environ Health*, vol. 15 (1967): 766–75. "No workers in the DDT plant developed cancer or blood dyscrasia in 19 years of continuous exposure up to 17 mg/man/day of DDT."

Laws, E.R. Regarding cancer research at Mayo Clinic: "Autopsies on all animals revealed no evidence of liver tumors or other hepatic pathology. The dosage is about 10,000 times that ingested by the average American in his daily diet."

In a March 15, 1978, hearing in San Francisco, more than seventy articles by cancer experts were discussed, specifying that DDT is not carcinogenic.

In 1964, Halver had trout developing tumors, and thought it might be caused by DDT in the water. In 1966, it was proven that the tumors were caused by aflatoxins in the cottonseed oil that was part of their diet. The aflatoxins, formed by Aspergillis mold, were in the fish food at concentrations of 6 parts per billion, and 1 part per billion is considered carcinogenic.

Reports by Tarjan, R. and Kemeny, T., 1966, were rejected after it was determined by WHO that all of the food given to their mice was contaminated with aflatoxins, causing even their "control mice" to develop tumors.

The Chief of Toxicology for the USPHS stated, "DDT is not a carcinogen," as did IARC director J. Higginson, at the EPA hearings.

Hart, L. and Fouts, J., *Arch Exp Path*, vol. 249 (1965): 486–520, DDT and its analogues stimulate production of hepatic microsomal drug-metabilizing enzymes. These enzymes are known to inhibit cancers.

DDT-induced enzymes detoxify alfatoxins, preventing the development of cancers or tumors in humans, birds, and other animals. This was discussed by Edward Laws (*Arch Environ Health* [Sept. 1971]: 23.), by Salinskas and Okey (*J Natl Cancer Inst*, vol. 55 [Sept. 1975]: 653–57.), by McLean and McLean (*British Med Bull*, vol. 25 (1969): 278–81), and many others.

Molnar, G.D., et al, *Cancer*, vol. 10: 259–68, Mayo Clinic. "Adrenal cancer patients benefited by DDT chemotherapy."

Okey, A.B. (*J Natl Cancer Inst*, vol. 55, no. 3: 653–57.) Leukemia development was inhibited by treatments with DDT.

Hazeltine, W. (*Clinical Toxicology*, vol. 4 (1971): 55–61.) "DDT analogue successfully treated juvenile jaundice."

Cancer was redefined by Russell Train, EPA administrator, in 1974. "For EPA's purposes, tumorigenic substances and carcinogenic substances are synonymous. Either 'tumors' or 'cancers' will henceforth be considered as 'cancer' by the EPA."

[The EPA could then invoke the "anti-cancer" DeLaney Clause of the FDA to justify banning chemicals which had never been known to cause any malignancies, and was supposed to deal with "cancer" rather than with non-malignant growths.]"

Index

About the Authors

Donald Roberts PhD.

Donald Roberts is a leading medical entomologist and academician researcher and Professor Emeritus of Tropical Public Health. He received his doctoral degree in 1973 from the University Of Texas School Of Public Health in Houston, Texas. He has worked around the world researching malaria and other vector-borne diseases. He was Chief, Department of Entomology, at the Walter Reed Army Institute of Research in Washington, D.C. and later was Professor of Tropical Public Health at the Uniformed Services University of the Health Sciences in Bethesda, MD. He is the author of over 120 peer-reviewed studies and has conducted considerable research on DDT and its role in public health. Don is also intensely invovled in advocacy to defend DDT and indoor residual spraying of public health insecticides for malaria control.

Richard Tren

Richard is an economist and co-founder and chairman of Africa Fighting Malaria (AFM), a malaria policy and advocacy group with offices in South Africa and Washington DC. For over a decade, Richard has been involved with issues of development and global health with a focus on malaria. Under Richard's leadership, AFM has successfully defended the use of DDT for malaria control and has advocated for more responsive, accountable and effective malaria control programs. Richard co-founded AFM during the negotiations of the Stockholm Convention and, by working with malaria scientists from around the world, helped to secure an exemption for DDT's continued use in malaria control.

Roger Bate

Roger Bate is the Legatum Fellow in Global Prosperity at the American Enterprise Institute. He researches international health policy, with a special interest in counterfeit medicines and malaria control. He has a PhD in economics from Cambridge University. Dr. Bate's writings have appeared in, among others, the *Wall Street Journal*, the *Financial Times*, the *Lancet, PLoS One, the Malaria Journal,* and the *British Medical Journal.* He regularly contributes to AEI's *Health Policy Outlook* series. He attended negotiating meetings in 1999 and 2000 in the build up to the UN Stockholm Convention on Persistent Organic Pollutants and argued against major restrictions on the use of DDT under the Convention.

Jennifer Zambone

Growing up in the formerly malarial swamps and marshes of New Jersey, Jennifer Zambone took an early and very personal interest in the mosquito and role in vector-borne disease. After receiving a bachelors of arts in biology and English and a juris doctorate from Washington and Lee University, Jennifer worked on regulatory issues before returning to her early interest in mosquitoes and their effects and joining the Africa Fighting Malaria as its Washington, DC director. She now works as the publications manager at the Mercatus Center at George Mason University.

CPSIA information can be obtained
at www.ICGtesting.com
Printed in the USA
BVHW03s0346010618
517874BV00008B/183/P

9 781608 443765